Fundamentals of Solid State Engineering

Fundamentals of Solid State Engineering

Editor

Hasad Demirci

scitus
academics

Fundamentals of Solid State Engineering

Edited by **Hasad Demirci**

Printed in 2017

ISBN: 978-1-68117-128-9
Library of Congress Control Number: 2015951997

© 2016 by
SCITUS Academics LLC,
616, Corporate Way, Suite 2, 4766,
Valley Cottage, NY 10989

www.scitusacademics.com

Notice

Reasonable efforts have been made to publish reliable data and views articulated in the chapters are those of the individual contributors, and not necessarily those of the editors or publishers. Editors or publishers are not responsible for the accuracy of the information in the published chapters or consequences of their use. The publisher believes no responsibility for any damage or grievance to the persons or property arising out of the use of any materials, instructions, methods or thoughts in the book. The editors and the publisher have attempted to trace the copyright holders of all material reproduced in this publication and apologize to copyright holders if permission has not been obtained. If any copyright holder has not been acknowledged, please write to us so we may rectify.

Preface

The century has witnessed the phenomenal rise of natural science and technology into all aspects of human life. Three major sciences have emerged and marked this century, physical science which has strived to understand the structure of atoms through quantum mechanics, life science which has attempted to understand the structure of cells and the mechanisms of life through biology and genetics, and information science which has symbiotically developed the communicative and computational means to advance natural science.Microelectronics has become one of today's principle enabling technologies supporting these three major sciences and touches every aspect of human life, such as food, energy, transportation, communication, entertainment, health and exploration.

This textbook presents the basic physics concepts and thorough treatment of semiconductor characterization technology, designed for solid state engineers.Thetext gives an overview of the basic multidisciplinary aspects of physical science. In the area of Solid State Physics in particular, it aims at teaching all the fundamental scientific concepts essential to solid state engineering. The book is primarily emphasized ina variety of fundamental scientific concepts essential to solid stateengineering, as well as the latest technological advances and modernapplications in this area.

Fundamentals of Solid State Engineering, delivers a multi-disciplinary introduction to solid state engineering, combining concepts from physics, chemistry, electrical engineering, materials science and mechanical engineering.

Table of Contents

CHAPTER 3   Comparison of Photoluminescence Properties of GD2O3
Phosphor Synthesized by Combustion and Solid State
Reaction Method... 5
Error! Bookmark not defined.

CHAPTER 4   Fabrication of Solid-state Thin-film Batteries using
LiMnPO4 thin films Deposited by Pulsed Laser
Deposition.. 7
Error! Bookmark not defined.

CHAPTER 1

Theories in Spin Dynamics of Solid-State Nuclear Magnetic Resonance Spectroscopy

Eugene S. Mananga[1,2*], Jalil Moghaddasi[1], Ajaz Sana[1] and Mostafa Sadoqi[2]

[1]Physics and Technology Department, BCC, CUNY, New York, USA
[2]Physics Department, St. John's University of New York City, New York, USA

ABSTRACT

This short review article presents theories used in solid-state nuclear magnetic resonance spectroscopy. Main theories used in NMR include the average Hamiltonian theory, the Floquet theory and the developing theories are the Fer expansion or the Floquet-Magnus expansion. These approaches provide solutions to the time-dependent Schrodinger equation which is a central problem in quantum physics in general and solid-state nuclear magnetic resonance in particular. Methods of these expansion schemes used as numerical integrators for solving the time dependent Schrodinger equation are presented. The action of their propagator operators is also presented. We highlight potential future theoretical and numerical directions such as the time propagation calculated by Chebychev expansion of the time evolution operators and an interesting transformation called the Cayley method.

INTRODUCTION

The Schrodinger equation is the fundamental equation of physics for describing quantum mechanical behavior. In classical physics, the Schrodinger equation predicts the future behavior of a dynamic system and plays an important role of Newton's laws and conservation of energy [1]. In quantum mechanics, the Schrodinger equation is a partial differential time dependent equation that describes how the quantum state of a physical system changes. The acceptability of Schrodinger equation lies on its applicability in various fields of sciences such as physics, chemistry, and materials science [2] - [6]. For instance in field such as nuclear magnetic resonance (NMR), much effort still needs to be done to explore several problems using the time-dependent Schrodinger equation. These problems include but are not limited to medical imaging, crystallography, ultra short strong laser pulses, biological systems, chemical structures and composition, spin dynamics of superconductors and semiconductors [7] - [26].

This short review presents some applications of major theories used in NMR spectroscopy such as the average Hamiltonian theory (AHT) and the Floquet theory (FLT), as well as the developing approaches including the Fer expansion (FE) and the Floquet-Magnus expansion (FME) [27] - [36] . We highlight potential future numerical and theoretical directions such as the time propagation operator calculated using Chebychev expansion and the transformation of Cayley [37] - [45]. The wealth of physical problems indicates the importance of having a general method for solving the time evolution of the density operator or the propagator operator in the case of NMR for instance. The density matrix $\rho(t)$ and its antecedent the propagator operator $U(t)$ have been extensively used for many-body systems, such as atoms, molecules and nuclei; polarization of light and angular correlation experiments; the theory of masers and maser-like devices; the mean field techniques, such as Hartree-Fock and Thomas-Fermi approximations; the description of atoms and molecules in strong electromagnetic fields; resonance fluorescence and resonance Raman in the presence of intense field. Vast applications are present in electron and nuclear magnetic resonances [6] [46] - [48]. The physical insights provided by the theories presented in this review are illustrated by their applications. The following schematic diagram (Figure 1) shows the Flow chart of the evolution operators, theories, foundations, numerical simulations and applications in NMR [47].

Solid-state NMR is a powerful method to elucidate molecular structure and dynamics in systems not amenable to characterization by other methodologies and its importance stands in its ability to accurately determine intermolecular distances and molecular torsion angles [34] [49] - [52].

Methods developed over the past 3.5 years enabled us to obtain simplified calculations for the common form of Hamiltonian in solid-state NMR and multimode Hamiltonian in its generalized Fourier expansion Hamiltonian [34] . Based on these and other unpublished findings, we now believe that the FME provides a quick and efficient means to calculate higher order terms allowing the disentanglement of the stroboscopic observation and effective Hamiltonian that will be useful to describe spin dynamics processes in solid-state NMR and understand different synchronized or non-synchronized experiments [34] [52] - [55].

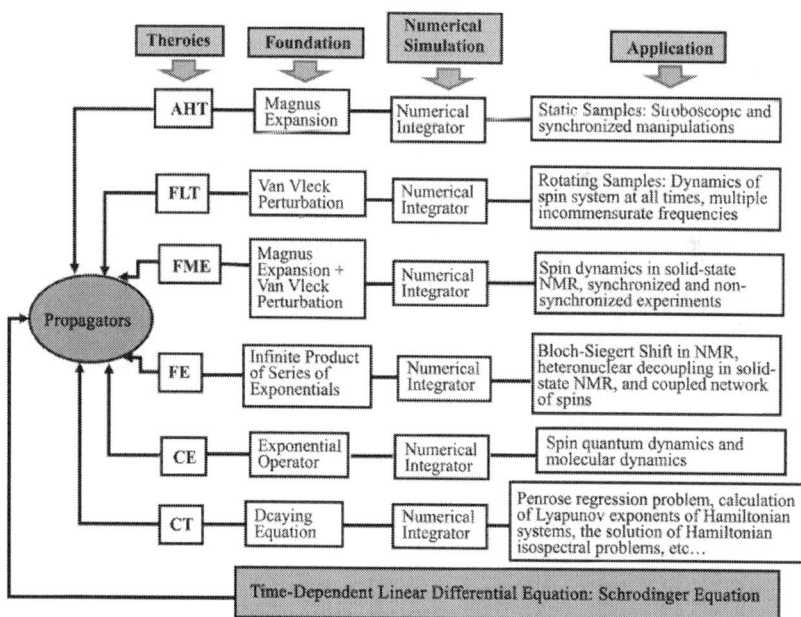

Figure 1. Flow chart of the evolution operators, theories, foundations, numerical simulations and applications in NMR.

Furthermore, our first applications of FE approach to study interactions in solid-state NMR when irradiated with the magic-echo sequence support this goal [52]. The results of the first order F_1 obtained for chemical shift, dipolar, and quadrupolar interactions might lead to the average

Hamiltonian, $\bar{H}$, in the sense of Magnus expansion under the circumstances:

$\frac{F_1(\tau_c)}{\tau_c} = \bar{H}$. A salient feature of the Fer and Magnus expansions stem from the fact that, when H(t) is an element in a given Lie algebra group, both approaches have the required structure and evolve in the desired group (Lie group). In addition, this is also true for their truncation to any order. We are thus poised to perform more work to ascertain the feasibility of Fer expansion in handling cases involving non-periodic and non-cyclic cases, and to use the expansion schemes of the Magnus (AHT) and the Fer expansions as numerical integrators for solving the time dependent Schrodinger equation which remains the central problem in quantum physics. Theoretical approaches in NMR are challenging, but the potential payoff is substantial, and could ultimately lead not only to a more accurate and efficient spin dynamics simulation, but also to the development of sophisticated RF pulse sequences, and understanding new experiments. Since the first demonstration of nuclear magnetic resonance in condensed matter in 1946 [56] [57], the field of solid-state NMR has adopted only two milestones theoretical approaches in its history, theories which control the dynamics of spin systems: the average Hamiltonian theory (1968) [27] and the Floquet theory (1982) [30] [31]. However, compared to other spectroscopic techniques, the technique of NMR is well-established and will remain much a vibrant field of research due to its theoretical components driven by mathematicians, chemical and quantum physicists.

The overall goals of this review article is to support theories in NMR in order to continue to a) apply the average Hamiltonian theory to problems including (but not limited to): a class of symmetrical radio-frequency pulse sequences in the NMR of rotating solids, the symmetry principles in the design of NMR multiple-pulse sequences, the composite pulses, and the problems still unsolved such as the AHT for 3 spins [58] - [68]; b) use the Floquet theory in the study of several magic-angle spinning (MAS) NMR experiments on spin systems with a periodically time-dependent Hamiltonian such as the multiple-multimode Floquet-theory in NMR [69]; c) enhance the performance of the Floquet-Magnus expansion by considering fundamental questions that arise when dealing with this approach [34]. Using FME method, many interesting problems will be approached such as multi-mode Hamiltonian, rotational-resonance recoupling, continuous wave irradiation on a single species, DARR and MIRROR recoupling, C-type and R-type sequences, TPPM decoupling, etc. [54] [68] [69] ; d) use the Fer expansion to solve similar problems such as those solved using the AHT [32] ; e) explore potential future theoretical and numerical directions for the calculation of the time propagation and evolution operators using Chebychev expansion and Cayley transformation methods [37] [41] - [44]. It is noteworthy that unifying or combining two and more theories known in NMR will continue to provide a framework for treating time-dependent Hamiltonian in quantum physics and NMR in a more efficient way that can be easily extended to all types of modulations.

AVERAGE HAMILTONIAN THEORY

Since its first application in NMR in 1968 by Evans, Haeberlen and Waugh, the average Hamiltonian theory has evolved as a powerful technique of analysis in the development of high resolution NMR spectroscopy [27] [70]. The Magnus expansion forms the basis of AHT and has been systematically used in NMR, in particular in solid-state NMR where via AHT the ME has been instrumental in the development of improved techniques in NMR spectroscopy [35] [36] [70]. The approach of AHT is the main tool to control the dynamics of spin systems and to treat theoretical problems in solid-state NMR which have been used sometimes abusively [71]. The basic understanding of AHT involves a time dependent Hamiltonian H(t) that governs the spin system evolution and describes the effective evolution by an average Hamiltonian $\bar{\bar{H}}$ within a periodic time t_c. This is satisfied only if H(t) is periodic t_c and the observation is stroboscopic and synchronized with period (t_c). This technique set the stage for stroboscopic manipulations of spins and spin interactions by radio-frequency pulses and also explains how periodic pulses can be used to transform the symmetry of selected interactions in coupled, many-spin systems considering the average or effective Hamiltonian of the RF pulse train [27] [34] [35] [72] - [75]. Wilhelm Magnus recognizes in his seminal paper of 1954 that his work was stimulated by results on the theory of linear operators in quantum mechanics. This shows that at its early stage, the Magnus expansion was strongly related to physics, and has been ever since then [72]. The central result of AHT is obtained by expressing the evolution propagator H(t$_c$) by an average Hamiltonian $\bar{H}_0$ and using the Magnus expansion. The Magnus expansion provides a solution to the initial value problem in terms of exponentials of combinations of the coefficient matrix $-iH(t)$.

$$\frac{dU(t)}{dt} = -iH(t)U(t), \ U(t_0)=U_0, \ t \in \Re, \ U(t) \in C^n, \ -iH(t) \in C^{n \times n},$$

(1)

The scalar case, n=1 (still valid for n>1 in some circumstances), has the general solution $U(t) = \exp\left(-\int_{t_0}^{t} iH(t')dt'\right)U_0$. If a term is added to the argument in the exponential such as $U(t) = \exp\left(-\int_{t_0}^{t} iH(t')dt' + M(t,t_0)\right)U_0$, then the Magnus expansion provides $M(t,t_0)$ as an infinite series. A salient feature of the Magnus expansion is the fact that, when $-iH(t)$ belong to a given Lie algebra, if we express $U(t)=U(t,t_0)U_0$, then $U(t,t_0)$ belong to the corresponding Lie group. By construction, the Magnus expansion lives in the Lie algebra. Furthermore, this is also true for their truncation to any order. In many applications this mathematical setting reflects important features of the problem. The method of AHT has been gradually applied to many theoretical problems in solid-state NMR such as pharmaceutical problem solving and methods development, symmetry in the design of NMR multiple- pulse sequences, composite pulses sequences, quantum computing, Magnus expansion as numericalintegrator, etc... [62] [65] [67] [72] [76] - [78]. Blanes and co-workers shown that the Magnus expansion can also be used as numerical method for solving Equation (1), with a good perspective of the overall performance of the numerical integrator provided by the efficiency diagram [35]. The efficiency plot is obtained by carrying out the numerical integrator with different time steps, corresponding to different numbers of evaluations of H(t). However, AHT is not applicable to Hamiltonians with multiple basic frequencies: MAS and radiofrequency irradiation must be synchronized or time-scale separated, multiple irradiations must be synchronized or time-scale separated [79] [80]. Our recent validation of the AHT method probed with quadrupolar nuclei showed that the AHT method becomes less efficient to predict the dynamics of the spin system as the quadrupolar spin nuclei dimension increase [75]. This is attributed to the Hilbert space becoming very large and leading to the contribution of non-negligible higher order terms in the Magnus expansion being truncated.

FLOQUET THEORY

The FLT introduced to the NMR community in the early 1980's simultaneously by Vegaand Maricqis another illuminating and powerful approach that offers a way to describe the time evolution of the spin system at all times and is able to handle multiple incommensurate frequencies [30] [31] [81] [82]. This theory provides a more general approach to AHT and has been applied satisfactorily to study important NMR phenomena [79] [80] [83]. The theory delineates the finite-dimensional time-dependent Liouville space onto an infinite-dimensional but time-independent Floquet space. The general description of the FLT is equally applicable to any nuclear spin systems. However, spin systems with large quadrupolar couplings may violate the convergence conditions for the expansions employed to evaluate the Floquet matrices. An important question to rise is the level of extension the FLT can be used in NMR without losing its conceptual framework. In other words, probing the validity of FLT for quadrupolar nuclei including those with spin I = 1, 3/2, 5/2, and 7/2 by analyzing for example a simple pulse sequence can be beneficial to the NMR community [75]. While the FLT scheme provides a more universal approach for the description of the full time dependence of the response of a periodically time-dependent system, it is most of the time impractical. Analytical calculations are limited to small spin systems and it is difficult to get physical insight from matrix representation. The full Floquet Hamiltonian has an infinite dimension and it is often not very intuitive to understand its implications on the time evolution of the spin system. Matrices for multi-mode Floquet calculations can become intractable. Massive reduction in dimensionality by truncation of the Fourier dimensions can introduce artifacts. In the literature, problems with up to three frequencies have been treated, but the demand of experiments that require four frequencies for a full description is increasing [69] [79] [80]. For instance, non-cyclic multiple-pulse sequences like two-pulse phase-modulated decoupling experiment acquire four frequencies under double rotation and there are some other obvious problems with four frequencies like triple-resonance CW radio frequency irradiation under MAS. Recent articles by Leskes et al., and Scholz et al. discussed extensively several MAS NMR experiments on spin systems with a periodically time-dependent Hamiltonian [69] [79] [80]. For many NMR experiments, understanding the spin dynamics requires a wise choice of the interaction frame in which the Hamiltonian is presented. Ramachandran and Griffin, and Schmidt and Vega introduced remarkable applications of Floquet theory in NMR [82] [83]. Indeed, bases employed in theoretical treatment of FLT and AHT do not extend to multiples spins or I > 1/2 systems, and fails to provide insights in to multiple-quantum NMR phenomena and polarization transfer experiments that involve relaxation. The multipole-multimode Floquet theory (MMFT) presented by Ramachandran and Griffin in its first application still remains a viable alternative for describing both coherent as well as incoherent effects observed in NMR experiments [83]. On one hand, Ramachandran and Griffin combined Shirley's Floquet approach to the multipole theory proposed by Sanctuary in order toexpand any periodic time-dependent spin Hamiltonian, density operator, and Liouville superoperator in a Fourier series [83] - [85]. Substituting the Fourier expansions of the density operator and the Liouville super-operator in the Liouville equation, the following new set of coupled differential equations spanning an infinite dimensional vector space, with time-independent coefficients were obtained

$$
i \frac{d\Phi(\bar{a},t)_{q,m_{1\rightarrow m}}^{(k)}}{dt} = \sum_{n'_{1\rightarrow m}=-\infty}^{\infty} \sum_{k',q',\alpha'} \sum_{l,m,\alpha_1} \left(Tr\left\{ T^{(k)-q}(\bar{a}) \times L\left(\overline{\alpha_1}\right)_{m_{1\rightarrow m}}^{lm} \cdot n'_{1\rightarrow m} \quad T^{(k')q'}\left(\overline{\alpha'}\right) \right\} + (n\cdot\omega)\delta \right) \Phi\left(\overline{\alpha'},t\right)_{q',n'_{1\rightarrow m}}^{(k')} .
$$

$$(2)$$

The notation $L\left(\overline{\alpha_1}\right)_{m_{1\rightarrow m}}^{(l)m} \cdot n'_{1\rightarrow m}$ includes the interaction coefficients as well as the spin and Fourier operators.

Subsequently, the Floquet density operator and the Hamiltonian operator are represented by

$$\rho_F\left(t\right) = \sum_{n_1 \to m = -\infty}^{\infty} \sum_{k,q,\bar{\alpha}} \Phi_{q,n_1 \to m}^{(k)}\left(\bar{\alpha},t\right) T^{(k)q}\left(\bar{\alpha}\right) F_{r_1}^1 F_{r_2}^2 L \ F_{r_m}^m,$$

(3)

$$H_F = \sum_{n_1 \to m = -\infty}^{\infty} \sum_{k,q,\bar{\alpha}} T^{(k)q}\left(\bar{\alpha}\right) F_{r_1}^1 F_{r_2}^2 L \ F_{r_m}^m + \sum_{i=1}^{m} \omega_i N^i.$$

(4)

The Floquet Hamiltonian H_F is represented using an operator basis constructed by the direct product of operators defined both in the spin $(T^{(k)q})$ as well as the Fourier dimensions ($F_{r_m}^m$, corresponding to m^{th} time modulation) with the off-diagonality represented by the indices q and r_m, respectively. This approach provides analytical insights in spite of the infinite dimensionality of the problem which can be validated by describing an analytical solution in the form of effective Hamiltonians obtained via contact or van Vleck transformation procedure [83]. On the other hand, Schmidt and Vega defined a set of Floquet operators that simplify the use of the Floquet theory for single spin system under MAS condition by considering the single spin system that exhibit a chemical shift MAS Hamiltonian defined in the spin state manifold $\{/\alpha>, /\beta>\}$ and the diagonalization of the Floquet Hamiltonian to the diagonalization of the sub-matrix diagonal matrices [82]. The signal and the Floquet transition amplitudes was evaluated to: $S(t) = e^{i\Delta\omega t}\sum_n I_n e^{in\omega_R t}$ with the I_n coefficients of $e^{i(n\omega_R t + \Delta\omega t)t}$ expressed as $I_n = \sum_m d_{m-n}^{\alpha\alpha} d_{-m}^{\alpha\alpha}$. Furthermore, both authors extended their investigation to the dipolar coupled I = 1/2 spin pairs by evaluating two uncoupled homonuclear spins under magic angle sample spinning conditions (i=1, 2......) with principal values of their chemical shift tensor $(\sigma_{11}^i, \sigma_{22}^i, \sigma_{33}^i)$ and Euler angles $(\alpha^i, \beta^i, \gamma^i)$ [82]. In this case, the Hamiltonian evaluated is represented by means of the operators X_n^{pq} which connect different Floquet states, namely $/p, m>$, with $/q, n+m>$ differing in the Fourier index n as well as the spin basis. Here, instead of calculating the correction terms, the method of contact transformation to calculate an effective Hamiltonian is used [86]. The contact transformation method is equivalent to the well-known Rayleigh-Schrodinger perturbation theory which provides corrections to zero order eigenvalues and eigenvectors. The unitary transformations are chosen in such a way that the off-diagonal operators due to interaction Hamiltonians are folded back to give diagonal contributions to the zero order Hamiltonian. As a result, a new Hamiltonian which is more effective, i.e., its eigenvalues are closer to the eigenvalues of the overall, untransformed Hamiltonian can be obtained. The transformation is done on the Hamiltonian so that by successive applications one obtains a Hamiltonian whose diagonal operators incorporate corrections from the interaction Hamiltonians [86]. The advantage of the method of contact transformation is that the correction is in the form of operators and therefore permits to define effective Hamiltonians which can be employed gainfully in pulse dynamics of rotating solids. An effective Hamiltonian is a simplified solution to the problem of finding the eigenvalues of the Floquet Hamiltonian. This method can also be useful when treating systems in which many spins are coupled. For instance, numerical diagonalization becomes quite difficult due to large dimensions of matrices when dealing with many spin coupled systems. Hence, the method of contact transformation gives the corrections in terms of operators and permits to restrict the spin basis, thereby reducing the size of matrices to be diagonalized in such systems.

FER EXPANSION

Analysis and numerical implementation of Magnus expansions is not a trivial task. Therefore, an alternative to the Magnus expansion which is called the Fer expansion can be useful for solving the time-dependent Schrodinger differential equation. This approach was formulated more than half a century ago by Fer and wasrecently introduced to the NMR community by Madhu and Kurur [32] [33]. This expansion is still in its infancy in NMR and can be considered to be complimentary to the Magnus expansion (AHT). Indeed, from the point of view of physical applications, the Magnus expansion has been extensively used in a variety of issues, while the Fer expansion has been either ignored or misquoted until recently [87]. While the efficiency of Fer expansion seems obvious, more effort is still required to allow the approach to overcome difficulties such as cases involving non- periodic and non-cyclic cases. More quantitative work need to be performed in order to bring out the salient features of the Fer expansion and explore its use in solid-state NMR and in many other theoretical areas. The Fer expansion approximates the solution to the initial value problem (Equation (1)) by a product of matrix exponentials. The expansion is generated by the recursive scheme,

$$U(t) = \prod_{k=1}^{\infty} e^{F_k(t)} = e^{F_1(t)} e^{F_2(t)} L \quad ,$$

and the iterative formula are

$$F_n = -i^n \int_0^t dt' H_F^{(n-1)}(t')$$

and

$$H_F^{(n)} = -\frac{1}{2}\left[F_n, H_F^{(n-1)}\right] + \frac{1}{3}\left[F_n,\left[F_n, H_F^{(n-1)}\right]\right] + L \quad ,$$

where $n = 1, 2, 3, L$.

The Fer expansion involves a series of nested commutators resulting in $H_F^{(n)}$. The Fer expansion differs to the Magnus approach in the form of the correction terms. The iteration process can continue easily when the initial values of $F_n(t)$ and $H_F^{(n)}$ are found. One major advantage of the Fer expansion over the AHT is that only an evaluation of nested commutators is required in the calculation of $H_F^{(n)}$. The Magnus expansion requires the calculation of nested commutators and their integrals to obtain the correction terms of a Hamiltonian. Blanes et al. had proved the convergence of the Fer expansion and showed that the convergence of Fer expansion is much faster than that of Magnus expansion [35] [36] [87]. Madhu and Kurar also highlighted the observations such that the calculation of a term like $H_F^{(1)}$ will contain several of the important signatures of the various higher- order terms in Magnus expansion, where all terms need to be calculated independently [32]. In addition, they mentioned that, the calculation of the infinite number of commutators, although looking imposing, may turn out to be simpler to handle in most experimentally interesting cases due to the fast convergence and the negligible value of many of the commutators. Both approaches (Fer and AHT) may be complimentary and the aspects of the problem at hand might eventually dictate the approach to be chosen [32]. The Fer expansion has been recently applied to the calculations of Block-Siegert shift in NMR, the analysis of heteronuclear decoupling in solid-state NMR, and the study of various interactions in solid-state NMR when irradiated with magic echo pulse sequence [32] [52]. Blanes and co-workers used Fer expansion as numerical method for solving time dependent Schrodinger equation. A good perspective of the overall performance of their given numerical integrator is provided by the efficiency diagram with the results better illustrated in a double logarithmic scale [35]. The Fer expansion has also been used to solve many physical situations such as classical time-dependent Hamiltonian systems [88] [89]. Furthermore, subtle aspects of

FE including, the convergence issue, the degree of computational involvement, and the application to coupled networks of spins, with regard to NMR still need to be tackle [32] [88].

FLOQUET-MAGNUS EXPANSION

The Floquet Magnus expansion is a new theoretical tool for describing spin dynamics recently introduced in solid-state NMR and spin physics [34] - [36]. This unique approach (FME) is an extension of the popular Magnus expansion and average Hamiltonian theory and is useful to shed new lights on AHT and FLT [27] [28]. The aims of the FME is to bridge the AHT to the Floquet Theorem but in a more concise and efficient formalism [34]. Calculations can then be performed in a finite-dimensional Hilbert space instead of an infinite dimensional space within the Floquet theory. We expected that the FME will provide means to more accurately and efficiently perform spin dynamics simulation and for devising new RF pulse sequence. We also expect the FME to explore physical implementations of quantum information processing (QIP) and introduce the basic background for understanding applications of NMR in QIP and explain their successes, limitations and potential. The FME provides a quick means to calculate higher order term allowing the disentanglement of the stroboscopic observation $\Lambda(t)$ and effective Hamiltonian F that will be useful to describe spin dynamics at all times in solid- state NMR and understand different synchronized or non-synchronized experiments. The FME offers a simple way to handle multiple incommensurate frequencies and thus open perspectives to deal with multi-mode Hamiltonian in the Hilbert space. This approach can provide new aspects not present in AHT and FT such as recursive expansion scheme in Hilbert space that can facilitate the development of new or improvement of existing pulse sequence. This scheme controls the spin dynamic systems in solid state NMR and makes use of its unique solution that has the required structure and evolves in the desired Lie group. In the first order, all three theoretical approaches (AHT, FLT, and FME) are equivalent, which corresponds to the popular average Hamiltonian, $H_{AHT}^{(0)} = H_{eff(FT)}^1 = H_{1(FME)} = H_0$. The FME approach can be considered as an improved AHT or a new version of FLT that could be very useful in simplifying calculations and providing a more intuitive understanding of spin dynamics processes. The approach of FME is essentially distinguished from other theories with its famous function $\Lambda_n(t)$ $(n=1,2,3,L)$ which provides an easy and alternative way for evaluating the spin behavior in between the stroboscopic observation points. The function $\Lambda_n(t)$ available only in the FME scheme will be useful to describe the spin dynamics in solid-state NMR and understanding different synchronized or non-syn- chronized experiments. The relationship with the regular Magnus expansion can be obtained from, $\frac{\Omega(T)}{T} = e^{-i\Lambda(0)} F e^{i\Lambda(0)}$ [34] - [36]. This points out that it is only in the case $\Lambda(0)=0$, that the FME gives the AHT as provided by the Magnus expansion, $\frac{\Omega(T)}{T} = F$. Therefore, the general approach of the AHT gives also the option of a more general representation of the FME with $\Lambda(0) \neq 0$. Furthermore, the function $\Lambda_n(t)$ is connected to the appearance of features like spinning sidebands in MAS. The general formulas for the contribution of the FME are given by:

$$\Lambda_n(t) = \Lambda_n(0) + \int_0^t G_n(x)\,dx - tF_n, \quad \text{with} \quad F_n = \frac{1}{T}\int_0^T G_n(x)\,dx$$

[34]. Symbolic calculation software can enable formal derivation of higher order terms. In the above equations, the $\Lambda_n(t)$ functions with $n=1,2,3,L$, re- presents the nth order term of the argument of the operator that introduces the frame such that the spin system operator is varying under the time independent Hamiltonian F. The evaluation of the function $\Lambda_n(t)$ is useful in many different ways, for instance, in rotating experiment of NMR, this function can be useful to quantify the level of productivity of double quantum terms [54] [55]. The FME propagator is given by: $U(t) = P(t)\exp\{-itH_F\}P^+(0)$. Here the constraint of stroboscopic observation is removed. $P(t)$ is the operator that introduces the frame that varies under the time independent Hamiltonian H_F. The function $\Lambda(t)$ given explicitly above is the argument of the operator $P(t)$ such that: $P(t) = \exp\{-i\Lambda t\}$. Like the FLT, the FME describes the time evolution of the spin system at all times. For various interactions in NMR, we recently calculated the first order function $(\Lambda_1(t))$ that provide an easy way for evaluating the spin system evolution [52]. The evaluation of $\Lambda_1(t)$ is useful especially for the analysis of the non-stroboscopic

evolution. We also found that the second order $^{(\Lambda_2(t))}$ is small in comparison to the first order$^{(\Lambda_1(t))}$, and will be less useful in many cases [54] [55].

Common Form of Hamiltonian in Solid-State NMR

For the sake of simplicity, we considered the Hamiltonian: $H = \omega_0 I_z + \lambda \sum_m (-1)^m R_{2,-m} T_{2,+m}$ which is a common form of Hamiltonian in solid-state NMR. $\omega_0 I_z$ is the Zeeman interaction, $R_{2,m}$ are the lattice parts of the internal interaction which encode its orientational dependence with respect to the magnetic field, $T_{2,m}$ are second rank m-order spherical tensor describing the spin system as defined by $[I_z, T_{2,m}] = m T_{2,m}$. The static perturbation theory (SPT) in terms of the irreducible tensor operators gives the diagonal Hamiltonian, $H_{SPT} = \omega_0 I_z + \lambda R_{2,0} T_{2,0} + \dfrac{\lambda^2}{2\omega_0} \sum_{m \neq 0} \dfrac{R_{2,m} R_{2,-m}}{m} [T_{2,m}, T_{2,-m}]$ [71]. Discrepancies between AHT and FT appear in the interaction frame where the Hamiltonian becomes time-dependent $\tilde{H}(t) = e^{i\omega_0 I_z t} H e^{-i\omega_0 I_z t} = \lambda \sum_m (-1)^m R_{2,-m} T_{2,m} e^{im\omega_0 t}$.
The FME provides an expansion in the rotating frame which is in agreement with the static perturbation theory and Van Vleck transformations. This is not the case of the Magnus expansion. This agreement can be easily explained by the connection that exists between the SPT and FME propagators written as $U_{SPT}(t) = e^{-iS} e^{-iH_{SPT} t} e^{iS}$, $U_{FME}(t) = \exp\left(-ie^{-i\omega_0 I_z t} \Lambda(t) e^{i\omega_0 I_z t}\right) e^{-i\omega_0 I_z t} e^{-iFt} e^{i\Lambda(0)}$, respectively. This means that under the criterion $S = \Lambda(0) = e^{-i\omega_0 I_z t} \Lambda(t) e^{i\omega_0 I_z t}$ both propagators describe the same evolution at any time [34].

Extension to Multimode Hamiltonian

Considering the generalized Fourier expansion of the Hamiltonian ($\omega = (\omega_1, \ldots, \omega_N)$ represented by the frequency indices) $H(t) = \sum_m H_m \exp(-im \cdot \omega t)$, we obtain $\Lambda_1(t) = \sum_{m \neq 0} \dfrac{H_m}{im \cdot \omega} e^{-im \cdot \omega t}$ and $F_1 = \sum_{m = 0} H_m$. Similarly, calculation of second order terms is straightforward [34]. These expressions highlight the fact that the multimode Hamiltonian case can be easily treated in Hilbert space with the FME.

BABA and C7

For example, applying the first contribution terms of FME to the dipolar Hamiltonian when irradiated with the BABA (Figure 2 and Figure 3) and sevenfold symmetric radiofrequency pulse sequences shown in Figure 3 of reference [54], we generated the plots ($\frac{A_1(t)}{b_{ij}\tau_R}$ and $\frac{A_2(t)}{b_{ij}^2\tau_R^2}$ versus the dimensionless numbers $\phi = \frac{t}{\tau_R}$) of the degree of recoupling magnetic dipolar between nuclear spins which is useful for preparing and detecting double quantum coherence [54] [55] [90]. Therefore, the study of the amplitude of DQ terms can be considered as a viable approach for controlling the complex spin dynamics of a spin system evolving under the dipolar interaction of BABA and C7 pulse sequences. The size of $\frac{A_1(t)}{b_{ij}\tau_R}$ determine the amplitude of the DQ coherence, which indicates the degree of efficiency of the scheme.

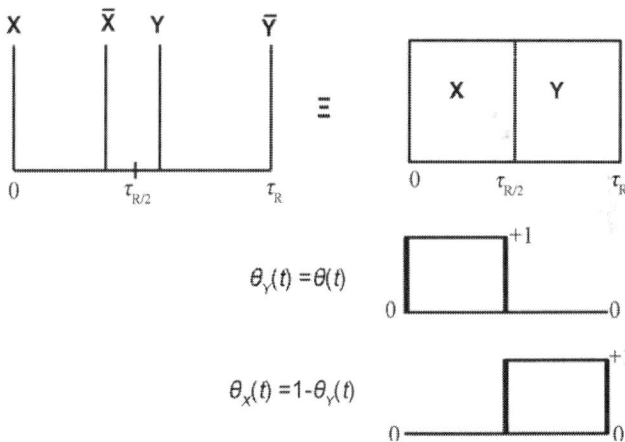

Figure 2. BABA with delta-pulses.

In reference [54] , a closer look at Figure 4(c) and Figure 5(c) shows that the magnitude of $\mathrm{BABA}_2(\phi)$ is small comparatively to the magnitude of $\mathrm{BABA}_1(\phi)$, i.e. $\frac{A_2(t)}{b_{ij}^2\tau_R^2} < \frac{A_1(t)}{b_{ij}\tau_R}$ as expected. As a result, $A_2(t)$ function will be less useful in many cases. We can also observe that all curves are strictly monotonous. This tells us that, the strength of the DQ terms increase continously with time and no decoupling conditions occur in the BABA (with delta-pulse) and C7 pulse sequences.

BABA with Finite Pulse Width

Now, let us Consider, BABA pulse sequence with finite pulse width where the relation $\theta_x(t) = 1 - \theta_y(t)$ is valid during the interval where $\theta(t)$ acted (Figure 3) [55] [90]. We investigated the simplest case and considered only DQ terms in the function $A_1(t)$. We generated two types of plots: $\frac{A_1(t)}{b_j \tau_R}$ versus the dimensionless numbers $\psi = \frac{t}{\tau_R}$ and $\phi = \frac{2\tau_p}{\tau_R}$ as shown in Figure 3 (b) and Figure 4(b) in reference [55]. We studied the case, $0.1 \le \phi \le 0.606$, corresponds to the spinning frequencies $\frac{\omega_R}{2\pi} = 5 - 10\,\text{kHz}$, and to the recoupling RF fields $\frac{\omega_{RF}}{2\pi} = 25 - 50\,\text{kHz}$. In reference [55], a closer look at Figure 3(b) (BABA with finite pulse widths) compared to BABA with delta-pulse width shows that the magnitude of the DQ terms of BABA with finite pulses is small compared to the magnitude of BABA with δ-pulse sequences, i.e. $\left.\left|\frac{A_1(t)}{b_j \tau_R}\right|\right|_{\text{finite-pulse}} < \left.\left|\frac{A_1(t)}{b_j \tau_R}\right|\right|_{\delta\text{-pulse}}$, as expected [55]. In reference [55], Figure 4(b) shows the plot of the function $\frac{A_1(t)}{b_j \tau_R}$ for versus the dimensionless number $\phi = \frac{2\tau_p}{\tau_R}$, for the two cases: $t = 1\,\text{ms}$ and $t = 2\,\text{ms}$.

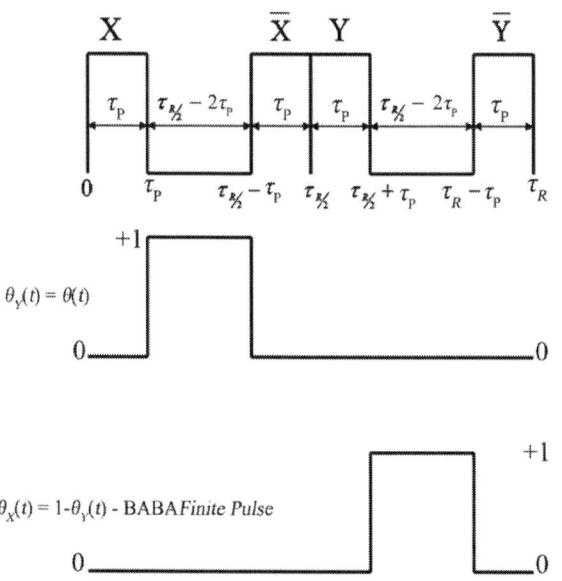

Figure 3. BABA with finite pulse width.

It can easily be seen that, when $\phi = \frac{2\tau_p}{\tau_R}$ increases, the magnitude of the double quantum terms decreases, as expected. When $\phi \to 0$, the magnitude of the DQ term ®
maximum corresponding to the delta-pulse sequence. However, when $\phi = 0.5$ corresponding to $\tau_p = \frac{\tau_R}{4}$, we have $\frac{\Lambda_1(t)}{b_{ij}} = 0$. The strength of the DQ terms decreases, cancel and build up again. This dynamic predicts that a full decoupling is possible, which occurs at $\phi = 0.5$. The plot of the magnitude of the double quantum term of $\Lambda_1(t)$ as a function of the pulse length gives a basic understanding of the experiment such as how to select robust finite pulse widths and how to select finite pulse widths that maximize or minimize double quantum terms. The study of this FME function could be helpful in predicting the conditions of decoupling.

Criteria to Average out Chemical Shift Anisotropy for BABA

Application of the first contribution terms of the Floquet-Magnus expansion to the chemical shift anisotropy when irradiated with the BABA pulse sequence lead to an important condition for the CSA to be averaged out in each rotor period τ_R [53] [91] . Considering the CSA interaction in the following general form, $H_{CSA}(t) = \sum_i \delta^i_{CSA}(t) I^i_z$, with $(ii = XX, YY, ZZ)$ $\delta^i_{CSA}(t) = \sum_{n=-2}^{+2} f^i_n(\alpha,\beta,\sigma^{ii}) \exp\{-in(\omega_r t + \gamma)\} R_{spin}(t)$, we obtained the criterion for the CSA to be averaged out in each τ_R period: $\frac{1}{\pi i}(f^i_1 - f^i_{-1}) + \frac{1}{2} f^i_0 = 0$ [91] . Similar criterion to average out CSA was obtained for BABA II (with $2\tau_R$), $\frac{1}{\pi i}(f^i_2 - f^i_{-2}) + \frac{1}{2} f^i_0 = 0$ [91]. The coefficients $f^i_n(\alpha,\beta,\sigma^{ii})$ depend on the orientation of the molecule and on the CSA tensor elements. The first order of the argument of the propagator operator in FME approach was evaluated to

$$\Lambda_1(t) = \frac{1}{2}\sum_i \sum_{n=-2}^{+2} f^i_n e^{-in\gamma}\left(\frac{-1}{in\omega_R}\right)(e^{-in\omega_R t}-1)(-I^i_X + I^i_Y)$$
$$+ \sum_i \sum_{m=-\infty}^{+\infty}\left(\frac{-1}{im\omega_R}\right)(e^{-im\omega_R t}-1)e^{-im\gamma}\sum_{n=-2}^{+2} f^i_n a_{m-n}(I^i_X + I^i_Y).$$

(5)

A numerical analysis for a simple case consisting of one spin system with $m=1$, $\gamma'=0$ evaluate instant neous values of the function $\frac{\Lambda_i(t)}{\tau_i}$. These values are the magnitude of the CSA in different orientation of the molecule and depend on the orientation of the molecule and on the CSA tensor elements. This complex function can also be ploted versus the dimensionless number $\phi=\frac{t}{\tau_i}$ to get insight of the magnitude of the CSA in different orientation of the molecule.

POTENTIAL APPROACHES AND FUTURE DIRECTIONS

Computing the exponential of a matrix is an important task in quantum mechanics and in nuclear magnetic resonance in particular where all theories used so far rely on exponential Hamiltonian operator propagators. The approximation of the matrix exponential is among the oldest and most extensive research topics in numerical mathematics [35] [39] [92] [93]. Although many efficient algorithms have been developed, so far, the problem is still not having being solved in general. Approaches such as scaling and squaring with Pade approximation, Chebyshev approximation, Krylov space methods, or splitting methods, have been used to approach the exponential of a matrix problem [39]. The main difficulty encountered in spectrum simulation is the rapid increase of computational requirements with an increasing number of spins. Simulation of spin system dynamics requires the numerical solution of the Liouville von Neumann equation, or equivalently the numerical exponential of a Liouville matrix [92] [93]. The Chebyshev approach has the potential to be extensively used in spin quantum dynamics and in particular NMR in the capacity of numerical simulations of spin physics of systems encounters. Our motivation of presenting the Chebyshev approximation as a potential surrogate of the popular expansions in NMR for the task of numerical simulations in spin dynamics paradigm stem from its numerical stability, high accuracy and also because its theoretical advantages are still not entirely realized for currently feasible computations [94] . I addition of the Chebyshev approximation, we introduce another alternative transformation called Cayley method that could be considered in some circumstances.

CHEBYSHEV APPROACH

Nearly three decades ago, Tal-Ezer and Kosloff introduced the Chebyshev method as a means of solving the time-dependent Schrodinger equation in the field of molecular dynamics [37] [92] [94]. Tal-Ezer shown that the complex Chebyshev polynomials achieve the best approximation to expand the evolution operator. In the Chebyshev approach, the evolution operator $\exp(-i\tau H)$ is expended in a truncated series of Chebyshev polynomials. This procedure is applied by bounding the extreme eigenvalues E_{min} and E_{max} of H. Then a truncated Chebyshev expansion of $\exp(-ix) \approx \sum_{n=0}^{m} c_n P_n(x)$ on the interval $[\tau E_{min}, \tau E_{max}]$ is considered where $P_n(x) = T_n\left(\dfrac{2x - \tau E_{max} - \tau E_{max}}{\tau E - \tau E}\right)$, with well-chosen coefficients c_n [35]. The Chebyshev method has two main advantages: first, it exploits the sparsity of the Liouvillian (Hamiltonian) by expressing the propagator in terms of a sequence of L (Liouville superoperator) matrix multiples. Second, the Chebyshev expansion of the propagator is essentially exact. The series converges so rapidly that it is easily extended to the point where the truncation error is smaller than the usual round-off errors expected in any numerical computation [94] [95]. The method of Chebyshev approximation is frequently used in numerical quantum dynamics to compute $\exp(-i\tau H)\psi_0$ over very long times. However, there are existing drawbacks in the Chebushev method. The scheme is not unitary, and therefore the norm is not conserved, but the deviation from unitarity is very small due to the extreme accuracy of the approach. Another drawback is that because of the long time durations of propagation in the Chebyshev scheme, intermediate results are not obtained.

CAYLEY METHOD

The Cayley transform provides a useful alternative to the exponential mapping relating the Lie algcbra to the Lie group. This fact is particularly important for numerical methods where the evaluation of the exponential matrix is the most computation-intensive part of the algorithm [35] [97].

Blanes and co-workers shown that the solution of Equation (1) can be written as $U(t)=\left(I-\frac{1}{2}C(t)\right)^{-1}\left(I+\frac{1}{2}C(t)\right)U_0$ with $C(t)$ satisfying the dcayinv equation, $\frac{dC}{dt}=A-\frac{1}{2}[C,A]-\frac{1}{4}CAC,$ and $t\geq t_0$, $C(t_0)=0$ [41] - [44]. A is element of the Lie algebra such that if B and C are also elements of a Lie algebra which can be combined by the Lie bracket, which we represent by $[A,B]=C$, with the consideration of the orthogonal group, the Calyley transform is $A=(I-\alpha B)^{-1}(I+\alpha B)$. Note that the choice of $\alpha=\frac{1}{2}$ is arbitrary but it ensures, a particular simple form of various expansion coefficients. Blanes and co-workers obtained the time-symmetric methods of order 4 and 6, based on the above Cayley transform where the efficiency of Cayley based methods can be built directly from Magnus based integrators [35]. But, unlike Magnus expansions, truncated Cayley expansions do not enjoy the benefits associated with time symmetry. As soon as integrals are replaced by appropriate quadrature formulas, Iserles proved that the time symmetry is gained [41]. The Cayley approach allows employing explicit schemes for solving the differential equation on the Lie algebra of the group and leads to semi-implicit methods where no iteration is required. The Caley methods in the numerical solution of matrix differential systems on quadratic groups have been applied to many important problems such as the Penrose regression problem (PRP) where this approach has been employed in finding numerical solution of PRP, the calculation of Lyapunov exponents of Hamiltonian systems, the solution of Hamiltonian isospectral problems, etc.... [41] -[44] [96].

CONCLUSIONS

In this publication, we have thoroughly reviewed the abiding applications of average Hamiltonian theory, Floquet theory, and Floquet-Magnus expansion from very different perpectives in spin quantum physics of nuclear magnetic resonance. We also have presented some potential theories in NMR such as Fer expansion, Chebychev approximation, and possibly Cayley method. The combinations of two or more of the theories therein described will provide a framework for treating time-dependent Hamiltonian in quantum physics and NMR in a way that can be easily extended to both synchronized and several non-synchronized modulations. We hope this publication will encourage the use of Floquet-Magnus and Fer expansions as numerical integrators as well as the use of Floquet-Magnus expansion as alternative approach in designing sophisticated pulse sequences and analyzing and understanding of different experiments. We also hope that this review will contribute to motivate spin dynamics experts in NMR to consider other perspectives and approaches beyond the scope of the current popular or used theories in the field of nuclear magnetic resonance. They are also many remarkable applications of the theory of NMR that we do not discuss in this review such as quantum information processing and computing. For example, the nuclear magnetic resonance quantum calculations of the Jones polynomial are interesting theoretical problems to tackle as well as theoretical treatment of problems with more than three frequencies analyzed using Floquet theory or Floquet-Magnus expansion approaches. In respect with the developments in the mathematical structure of AHT, FLT, FME, and FE, we expect that the realm of applications of the Floquet Magnus expansion and Fer expansion will also wide over the years. With new application in the field of NMR, we also expect the FME to generate new contributions like the generation of efficient numerical algorithm for geometric integrators.

The intention of writing this overview of theories and applications in nuclear magnetic resonance spectroscopy is to help bring the current and future prospective theoretical aspects of spin dynamics in NMR to the attention of the NMR community and lead new interactions between NMR experts and other specialists in mathematics, physics, chemistry, physical chemistry, and chemical physics. All these points strongly support the idea that the Floquet-Magnus expansion, the Fer expansion, the Chebyshev approach, and possibly the Cayley method can also be the very useful and powerful tools in quantum spin dynamics.

ACKNOWLEDGEMENTS

E. S. Mananga appreciates the moral supports of Profs. Joseph Malinsky, Andrew Akinmoladun and Akhil Lal, Mr. Hamad Khan and Mr. Alfred Romito.

REFERENCES

1. Schrödinger, E. (1926) An Undulatory Theory of the Mechanics of Atoms and Molecules. Physical Review, 28, 1049- 1970. http://dx.doi.org/10.1103/PhysRev.28.1049

2. Dirac, P.A.M. (1958) The Principles of Quantum Mechanics. 4th Edition, Oxford University Press, Oxford.

3. Hazewinkel, M. (2001) Schrodinger Equation. Encyclopedia of Mathematics, Edition Springer.

4. Müller-Kirsten, H.J.W. (2012) Introduction to Quantum Mechanics: Schrödinger Equation and Path Integral. 2nd Edition, World Scientific. http://dx.doi.org/10.1142/8428

5. Griffiths, D.J. (2004) Introduction to Quantum Mechanics. 2nd Edition, Benjamin Cummings.

6. Berman, M. and Kosloff, R. (1991) Time-Dependent Solution of the Liouville-Von Neumann Equation: Non-Dissipa- tive Evolution. Computer Physics Communications, 63, 1-20. http://dx.doi.org/10.1016/0010-4655(91)90233-B

7. Andrew, E.R., Bradbury, A. and Eades, R.G. (1958) Nuclear Magnetic Resonance Spectra from a Crystal Rotated at High Speed. Nature, 182, 1659.http://dx.doi.org/10.1038/1821659a0

8. Andrew, E.R., Bradbury, A. and Eades, R.G. (1959) Removal of Dipolar Broadening of Nuclear Magnetic Resonance Spectra of Solids by Specimen Rotation. Nature, 183, 1802-1803. http://dx.doi.org/10.1038/1831802a0

9. Lowe, I.J. (1959) Free Induction Decays of Rotating Solids. Physical Review Letters, 2, 285-287. http://dx.doi.org/10.1103/PhysRevLett.2.285

10. Schaefer, J. and Stejskal, E.O. (1976) Carbon-13 Nuclear Magnetic Resonance of Polymers Spinning at the Magic Angle. Journal of American Chemical Society, 98, 1031.http://dx.doi.org/10.1021/ja00420a036

11. Ernst, R.R., Bodenhausen, G. and Wokaun, A. (1987) Principles of Nuclear Magnetic Resonance in One and Two Dimensions. Clarendon, Oxford.

12. Hafner, S. and Demco, D.E. (2002) Solid-State NMR Spectroscopy under Periodic Modulation by Fast Magic Angle Spinning and Pulses: A Review. Solid State Nuclear Magnetic Resonance, 22, 247-274.

13. Mehring, M., Pines, A., Rhim, W.-K. and Waugh, J.S. (1971) Spin-Decoupling in the Resolution of Chemical Shifts in Solids by Pulsed NMR. Journal of Chemical Physics, 54, 3239-3240. http://dx.doi.org/10.1006/snmr.2002.0088

14. Jaroniec, C.P., Tounge, B.A., Rienstra, C.M., Herzfeld, J. and Griffin, R.G. (2000) Recoupling of Heteronuclear Dipolar Interactions with Rotational-Echo Double-Resonance at High Magic-Angle Spinning Frequencies. Journal of Magnetic Resonance, 146, 132-139.http://dx.doi.org/10 .1006/jmre.2000.2128

15. Charpentier, T., Fermon, C. and Virlet, J. (1998) Efficient Time Propagation Technique for MAS NMR Simulation: Application to Quadrupolar Nuclei. Journal of Magnetic Resonance, 132, 181-190. http://dx.doi.org/10.1006/jmre.1998.1415

16. Ernst, M., Geen, H. and Meier, B.H. (2006) Amplitude-Modulated Decoupling in Rotating Solids: A Bimodal Floquet Approach. Solid State Nucleaire Magnetic Resonance, 29, 2-21.http://dx.doi.org/10.1016/j.ssnmr.2005.08.004

17. Barone, S.R., Narcowich, M.A. and Narcowich, F.J. (1977) Floquet Theory and Applications. Physical Review A, 15, 1109-1125. http://dx.doi.org/10.1103/PhysRevA.15.1109

18. Filip, C., Filip, X., Demco, D.E. and Hafner, S. (1997) Spin Dynamics under Magic Angle Spinning by Floquet Formalism. Molecular Physics, 92, 757-771.http://dx.doi.org/10.1080/002689797170031

19. Friedrich, U., Schnell, I., Brown, S.P., Lupulescu, A., Demco, D.E. and Spiess, H.W. (1998) Spinning-Sideband Patterns in Multiple-Quantum Magic-Angle Spinning NMR Spectroscopy. Molecular Physics, 95, 1209-1227.http://dx.doi.org/10.1080/00268979809483252

20. Boender, G.J., Vega, S. and De Groot, H.J.M. (2000) Quantized Field Description of Rotor Frequency-Driven Dipolar Recoupling. Journal of Chemical Physics, 112, 1096-1106.http://dx.doi.org/10.1063/1.480664

21. Ernst, M., Samoson, A. and Meier, B.H. (2005) Decoupling and Recoupling Using Continuous Waves Irradiation in Magic-Angle-Spinning Solid-State NMR: A Unified Description Using Bimodal Floquet Theory. Journal of Chemical Physics, 123, Article ID: 064102. http://dx.doi.org/10.1063/1.1944291

22. Ding, S. and McDowell, C.A. (1998) The Equivalence between Floquet Formalism and the Multi-Step Approach in Computing the Evolution Operator of a Periodical Time-Dependent Hamiltonian. Chemical Physics Letters, 288, 230- 234. http://dx.doi.org/10.1016/S0009-2614(98)00307-8

23. Buishvili, L.L., Volzhan, E.B. and Menabde, M.G. (1981) Higher Approximations in the Theory of the Average Hamiltonian. Theoretical and Mathematical Physics, 46, 166-173.http://dx.doi.org/10.1007/BF01030852

24. Emetere, M.E. (2014) Analytical Solutions of Three Dimensional Time-Dependent Shrodinger Equation Using Bloch NMR Approach for NMR Studies. Applied Mathematical Sciences, 8, 2753-2762. http://dx.doi.org/10.12988/ams.2014.4012

25. Lee, Y.K., Kurur, N.D., Helmle, M., Johannessen, O.G., Nielsen, N.C. and Levitt, M.H. (1995) Efficient Dipolar Recoupling in the NMR of Rotating Solids. A Sevenfold Symmetric Radiofrequency Pulse Sequence. Chemical Physical Letters, 242, 304-309.http://dx.doi.org/10.1016/0009-2614(95)00741-L

26. Leforestier, C., Bisseling, R.H., Cerjan, C., Feit, M.D., Friesner, R., Guldberg, A., Hammerich, A., Jolicard, G., Karrlein, W., Meyer, H.-D., Lipkin, N., Roncero, O. and Kosloff, R. (1991) A Comparison of Different Propagation Schemes for the Time Dependent Schrödinger Equation. Journal of Computational Physics, 94, 59-80.http://dx.doi.org/10.1016/0021-9991(91)90137-A

27. Haeberlen, U. and Waugh, J.S. (1968) Coherent Averaging Effects in Magnetic Resonance. Physical Review, 175, 453- 467. http://dx.doi.org/10.1103/PhysRev.175.453

28. Floquet, G. (1883) Sur les Equations Differentielles Lineaires a Coefficients Periodiques. Annales Scientifics de l'Ecole Normale Superieur, 12, 47-88. http://eudml.org/doc/80895

29. Shirley, J.H. (1965) Solution of the Schrödinger Equation with a Hamiltonian Periodic in Time. Physical Review, 138, B979-B987. http://dx.doi.org/10.1103/PhysRev.138.B979

30. Maricq, M.M. (1982) Application of Average Hamiltonian Theory to the NMR of Solids. Physical Review B25, 6622-6632. http://dx.doi.org/10.1103/PhysRevB.25.6622

31. Zur, Y., Levitt, M.H. and Vega, S. (1983) Multiphoton NMR Spectroscopy on a Spin System with I=1/2. Journal of Chemical Physics, 78, 5293.http://dx.doi.org/10.1063/1.445483

32. Madhu, P.K. and Kurur, N.D. (2006) Fer Expansion for Effective Propagators and Hamiltonians in NMR. Chemical Physical Letters, 418, 235-238.http://dx.doi.org/10.1016/j.cplett.2005.10.134

33. Fer, F. (1958) Bulletin de la Classe des Sciences. Academie Royalede Belgique, 44, 818-829.

34. Mananga, E.S. and Charpentier, T. (2011) Introduction of the Floquet-Magnus Expansion in Solid-State Nuclear Magnetic Resonance Spectroscopy. Journal of Chemical Physics, 135, Article ID: 044109. http://dx.doi.org/10.1063/1.3610943

35. Blanes, S., Casas, F., Oteo, J.A. and Ros, J. (2009) The Magnus Expansion and Some of Its Applications. Physics Reports, 470, 151-238.http://dx.doi.org/10.1016/j.physrep.2008.11.001

36. Casas, F., Oteo, J.A. and Ros, J. (2001) Floquet Theory: Exponential Perturbative Treatment. Journal of Physics A: Mathematical and General, 34, 3379-3388.http://dx.doi.org/10.1088/0305-4470/34/16/305

37. Dumont, R.S., Jain, S. and Bain, A. (1997) Simulation of Many-Spin System Dynamics via Sparse Matrix Methodology. Journal of Chemical Physics, 106, 5928.http://dx.doi.org/10.1063/1.473258

38. Scholz, I., Meier, B.H. and Ernst, M. (2007) Operator-Based Triple-Mode Floquet Theory in Solid-State NMR. Journal of Chemical Physics, 127, Article ID: 204504.http://dx.doi.org/10.1063/1.2800319

39. Süli, E. and Mayers, D. (2003) An Introduction to Numerical Analysis. Cambridge University Press, Cambridge.

40. Abramowitz, M. and Stegun, I.A. (1965) Handbook of Mathematical Functions. Dover Publications, Dover.

41. Iserles, A. (2001) On Cayley-Transform Methods for the Discretization of Lie-Group Equations. Foundations on Computational Mathematics, 1, 129-160.http://dx.doi.org/10.1007/s102080010003

42. Chu, M.T. and Morris, L.K. (1988) Isospectral Flows and Abstract Matrix Factorization. Society for Industrial and Applied Mathematics Journal of Numerical Analysis, 25, 1383-1391. http://dx.doi.org/10.1137/0725080

43. Leimkuhler, J.B. and Van Vleck, E.S. (1997) Orthosymplectic Integration of Linear Hamiltonian System. Numerical Mathematics, 77, 269-282.http://dx.doi.org/10.1007/s002110050286

44. Lewis, D. and Simo, J.C. (1994) Conserving Algorithms for the Dynamics of Hamiltonian Systems on Lie Groups. Journal of Nonlinear Science, 4, 253-299.http://dx.doi.org/10.1007/BF02430634

45. Abragam, A. (1961) The Principle of Nuclear Magnetism. Clarendon Press, Oxford.

46. Ben-Reuven, A. and Rabin, Y. (1979) Theory of Resonance Scattering and Absorption of Strong Coherent Radiation by Thermally Relaxing Multi-Level Atomic Systems. Physical Review A, 19, 2056-2073. http://dx.doi.org/10.1103/PhysRevA.19.2056

47. Mananga, E.S. (2014) Future Theoretical Approaches in Nuclear Magnetic Resonance. Journal of Modern Physics, 5, 145-148.

48. Raleigh, D.P., Levitt, M.H. and Griffin, R.G. (1988) Rotational Resonance in Solid State NMR. Chemical Physics Letters, 146, 71-76. http://dx.doi.org/10.1016/0009-2614(88)85051-6

49. Hohwy, M., Rienstra, C.M., Jaroniec, C.P. and Griffin, R.G. (1999) Fivefold Symmetric Homonuclear Dipolar Recoupling in Rotating Solids: Application to Double Quantum Spectroscopy. Journal of Chemical Physics, 110, 7983.http://dx.doi.org/10.1063/1.478702

50. Ishii, Y., Terao, T. and Kainosho, M. (1996) Relayed Anisotropy Correlation NMR: Determination of Dihedral Angles in Solids. Chemical Physics Letters, 265, 133-140.http://dx.doi.org/10.1016/0009-2614(96)00426-5

51. Ishii, Y., Hirao, K., Terao, T., Terauchi, T., Oba, M., Nishiyama, K. and Kainosho, M. (1998) Determination of Peptide Angles in Solids by Relayed Anisotropy Correlation NMR. Solid State Nuclear Magnetic Resonance, 11, 169- 175. http://dx.doi.org/10.1016/S0926-2040(98)00038-1

52. Mananga, E.S. (2013) Applications of Floquet-Magnus Expansion, Average Hamiltonian Theory and Fer Expansion to Study Interactions in Solid State NMR When Irradiated with the Magic-Echo Sequence. Solid State Nuclear Magnetic Resonance, 55-56, 54-62.http://dx.doi.org/ 10.1016/ j.ssnmr.2013.08.002

53. Mananga, E.S. (2013) Criteria to Average Out the Chemical Shift Anisotropy in Solid-State NMR When Irradiated with BABA I, BABA II, and C7 Radiofrequency Pulse Sequences. Solid State Nuclear Magnetic Resonance, 55-56, 63-72.http://dx.doi.org/10.1016/j.ssnmr.2013.08.003

54. Mananga, E.S., Reid, A.E. and Charpentier, T. (2012) Efficient Theory of Dipolar Recoupling in Solid-State Nuclear Magnetic Resonance of Rotating Solids using Floquet-Magnus Expansion: Application on BABA and C7 Radiofrequency Pulse Sequences. Solid State Nuclear Magnetic Resonance, 41, 32-47.http://dx.doi.org/10.1016/j.ssnmr.2011.11.004

55. Mananga, E.S. and Reid, A.E. (2013) Investigation of the Effect of Finite Pulse Errors on the BABA Pulse Sequence using the Floquet-Magnus Expansion Approach. Molecular Physics, 111, 243-257. http://dx.doi.org/ 10.1080/00268976.2012.718379

56. Purcell, E.M., Torrey, H.C. and Pound, R.V. (1946) Resonance Absorption by Nuclear Magnetic Moments in a Solid. Physical Review, 69, 37-38.http://dx.doi.org/10.1103/PhysRev.69.37

57. Bloch, F., Hansen, W.W. and Packard, M. (1946) The Nuclear Induction Experiment. Physical Review, 70, 474-485. http://dx.doi.org/ 10.1103/PhysRev.70.474

58. Brinkmann, A., Eden, M. and Levitt, M.H. (2000) Synchronous Helical Pulse Sequences in Magic-Angle Spinning Nuclear Magnetic Resonance: Double Quantum Recoupling of Multiple-Spin Systems. Journal of Chemical Physics, 112, 8539.http://dx.doi.org/10.1063/1.481458

59. Hohwy, M., Jakobsen, H.J., Eden, M., Levitt, M.H. and Nielsen, N.C. (1998) Broadband Dipolar Recoupling in the Nuclear Magnetic Resonance of Rotating Solids: A Compensated C7 Pulse Sequence. Journal of Chemical Physics, 108, 2686.http://dx.doi.org/10.1063/1.475661

60. Brinkmann, A. and Levitt, M.H. (2001) Symmetry Principles in the Nuclear Magnetic Resonance of Spinning Solids: Heteronuclear Recoupling by Generalized Hatmann-Hahn Sequences. Journal of Chemical Physics, 115, 357. http://dx.doi.org/10.1063/1.1377031

61. Carravetta, M., Eden, M., Zhao, X., Brinkmann, A. and Levitt, M.H. (2000) Symmetry Principles for the Design of Radiofrequency Pulse Sequences in the Nuclear Magnetic Resonance of Rotating Solids. Chemical Physical Letters, 321, 205-215.http://dx.doi.org/10.1016/S0009-2614(00)00340-7

62. Tycko, R. (2007) Symmetry-Based Constant-Time Homonuclear Dipolar Recoupling in Solid-State NMR. Journal of Chemical Physics, 126, Article ID: 064506.http://dx.doi.org/10.1063/1.2437194

63. Tycko, R. (2008) Lecture Notes for the First Winter School on Biomolecular Solid State NMR, Stowe, Vermont, 20-25 January 2008.

64. Eden, M. and Levitt, M.H. (1999) Pulse Sequence Symmetries in Nuclear Magnetic Resonance of Spinning Solids: Application to Heteronuclear Decoupling. Journal of Chemical Physics, 111, 1511. http://dx.doi.org/10.1063/1.479410

65. Sanctuary, B.C. and Cole, H.B.R. (1987) Multipole Theory of Composite Pulses. Journal of Magnetic Resonance, 71, 106-115. http://dx.doi.org/10.1016/0022-2364(87)90131-4

66. Levitt, M.H. (1982) Symmetrical Composite Pulse Sequences for NMR Population Inversion. I. Compensation of Radiofrequency Field Inhomogeneity. Journal of Magnetic Resonance, 48, 234-264. http://dx.doi.org/10.1016/ 0022-2364(82)90275-X

67. Tycko, R., Cho, H.M., Schneider, E. and Pines, A. (1985) Composite Pulses without Phase Distortion. Journal of Magnetic Resonance, 61, 90-101. http://dx.doi.org/10.1016/0022-2364(85)90270-7

68. Leskes, M., Madhu, P.K. and Vega, S. (2010) Floquet Theory in Solid-State Nuclear Magnetic Resonance. Progress in Nuclear Magnetic Resonance Spectroscopy, 55, 345-380.http://dx.doi.org/10.1016/j.pnmrs.2010.06.002

69. Scholz, I., Van Beek, J.D. and Ernst, M. (2010) Operator-Based Floquet Theory in Solid-State NMR. Solid State Nuclear Magnetic Resonance, 37, 39-59.http://dx.doi.org/10.1016/j.ssnmr.2010.04.003

70. Evans, W. (1968) On Some Applications of Magnus Expansion in Nuclear Magnetic Resonance. Annals of Physics, 48, 72-93. http://dx.doi.org/10.1016/0003-4916(68)90270-4

71. Goldman, M., Grandinetti, P.J., Llor, A., Olejniczak, Z., Sachleben, J.R. and Zwanziger, J.W. (1992) Theoretical Aspects of Higher-Order Truncations in Solid-State Nuclear Magnetic Resonance. Journal of Chemical Physics, 97, 8947.http://dx.doi.org/10.1063/1.463321

72. Tycko, R. (2008) Introduction to Special Topic: New Developments in Magnetic Resonance. Journal of Chemical Physical, 128, Article ID: 052101.http://dx.doi.org/10.1063/1.2833958

73. Mananga, E.S., Roopchand, R., Rumala, Y.S. and Boutis, G.S. (2007) On the Application of Magic Echo Cycles for Quadrupolar Echo Spectroscopy of Spin-1 Nuclei. Journal of Magnetic Resonance, 185, 28-37. http://dx.doi.org/10.1016/j.jmr.2006.10.016

74. Mananga, E.S., Rumala, Y.S. and Boutis, G.S. (2006) Finite Pulse Width Artifact Suppression in Spin-1 Quadrupolar Echo Spectra by Phase Cycling. Journal of Magnetic Resonance, 181, 296-303. http://dx.doi.org/10.1016/j.jmr.2006.05.015

75. Mananga, E.S., Hsu, C.D., Ishmael, S., Islam, T. and Boutis, G.S. (2008) Probing the Validity of Average Hamiltonian Theory for Spin I=1, 3/2 and 5/2 Nuclei by Analyzing a Simple Two-Pulse Sequence. Journal of Magnetic Resonance, 193, 10-22.http://dx.doi.org/10.1016/j.jmr.2008.03.014

76. Levitt, M.H. (2008) Symmetry in the Design of NMR Multiple-Pulse Sequences. Journal of Chemical Physics, 128, Article ID: 052205. http://dx.doi.org/10.1063/1.2833958

77. Vandersypen, L.M.K. and Chuang, I.L. (2004) NMR Techniques for Quantum Control and Computation. Review of Modern Physics, 76, 1037-1069.http://dx.doi.org/10.1103/RevModPhys.76.1037

78. Hu, B., Delevoye, L., Lafon, O., Trebosc, J. and Amoureux, J.P. (2009) Double-Quantum NMR Spectroscopy of 31P Species Submitted to Very Large CSAs. Journal of MagneticResonance, 200, 178-188. http://dx.doi.org/10.1016/j.jmr.2009.06.020

79. Scholz, I., Van Beek, J.D. and Ernst, M. (2010) Operator-Based Floquet Theory in Solid-State NMR. Solid State Nuclear Magnetic Resonance, 37, 39-59.http://dx.doi.org/10.1016/j.ssnmr.2010.04.003

80. Leskes, M., Akbey, U., Oschkinat, H., van Rossum, B.-J. and Vega, S. (2011) Radio Frequency Assisted Homonuclear Recoupling—A Floquet Description of Homonuclear Recoupling via Surrounding Heteronuclei in Fully Protonated to Fully Deuterated Systems. Journal of Magnetic Resonance, 209, 207-219.http://dx.doi.org/10.1016/j.jmr.2011.01.015

81. Mehring, M. (1983) Principles of High Resolution NMR in Solids. Springer-Verlag, New York.

82. Schmidt-Rohr, K. and Spiess, H.W. (1996) Multidimensional Solid-State NMR and Polymers. Academic Press, London.

83. Schmidt, A. and Vega, S. (1992) The Floquet Theory of Nuclear Magnetic Resonance Spectroscopy of Single Spins and Dipolar Coupled Spin Pairs in Rotating Solids. Journal of Chemical Physics, 96, 2655. http://dx.doi.org/10.1063/1.462015

84. Ramachandran, R. and Griffin, R.G. (2005) Multipole-Multimode Floquet Theory in Nuclear Magnetic Resonance. Journal of Chemical Physics, 122, Article ID: 164502.http://dx.doi.org/10.1063/1.1875092

85. Sanctuary, B.C. (1976) Multipole Operators for an Arbitrary Number of Spins. Journal of Chemical Physics, 64, 4352. http://dx.doi.org/10.1063/1.432104

86. Sanctuary, B.C. (1983) Multipole NMR. Molecular Physics, 48, 1155-1176.http://dx.doi.org/10.1080/00268978300100841

87. Ramesh, R. and Krishnan, M.S. (2001) Effective Hamiltonians in Floquet Theory of Magic Angle Spinning Using van Vleck Transformation. Journal of Chemical Physics, 114, 5967.http://dx.doi.org/10.1063/1.1354147

88. Blanes, S., Casas, F., Oteo, J.A. and Ros, J. (1998) Magnus and Fer Expansions for Matrix Differential Equations: The Convergence Problem. Journal of Physics A: Mathematical and General, 31, 259-268. http://dx.doi.org/10.1088/0305-4470/31/1/023

89. Haeberlen, U. (1976) Advance in Magnetic Resonance. Supplement 1, Academic Press, New York.

90. Casa, F., Oteo, J.A. and Ros, J. (1991) Lie Algebraic Approach to Fer's Expansion for Classical Hamiltonian Systems. Journal of Physics A: Mathematical and General, 24, 4037-4046. http://dx.doi.org/10.1088/0305-4470/24/17/020

91. Feike, M., Demco, D.E., Graf, R., Gottwald, J., Hafner, S. and Spiess, H.W. (1996) Broadband Multiple-Quantum NMR Spectroscopy. Journal of Magnetic Resonance A, 122, 214-221. http://dx.doi.org/ 10.1006/jmra.1996.0197

92. Mananga, E.S. (2013) Progress in Spin Dynamics Solid-State Nuclear Magnetic Resonance with the Application of Floquet-Magnus Expansion to Chemical Shift Anisotropy. Solid State Nuclear Magnetic Resonance, 54, 1-7.http://dx.doi.org/10.1016/j.ssnmr.2013.04.001

93. Tal-Ezer, H. and Kosloff, R. (1984) An Accurate and Efficient Scheme for Propagating the Time Dependent Schrödinger Equation. Journal of Chemical Physics, 81, 3967.http://dx.doi.org/10.1063/1.448136

94. Moler, C.B. and Van Loan, C.F. (2003) Nineteen Dubious Ways to Compute the Exponential of a Matrix, Twenty-Five Years Later. Society for Industrial and Applied Mathematics Review, 45, 3-49. http://dx.doi.org/10.1137/S00361445024180

95. Rivlin, T.J. (1990) Chebychev Polynomials. 2nd Edition, Wiley, New York.

96. Blanes, S., Casas, F. and Ros, J. (2002) High Order Optimized Geometric Integrators for Linear Differential Equations. BIT, 42, 262-284.

97. Lopez, L. and Politi, T. (2001) Applications of the Cayley Approach in the Numerical Solution of Matrix Differential Systems on Quadratic Groups. Applied Numerical Mathematics, 36, 35-55. http://dx.doi.org/10.1016/S0168-9274(99)00049-5

CITATION

Mananga, E. , Moghaddasi, J. , Sana, A. and Sadoqi, M. (2015) Theories in Spin Dynamics of Solid-State Nuclear Magnetic Resonance Spectroscopy. World Journal of Nuclear Science and Technology, 5, 27-42. doi: 10.4236/wjnst.2015.51004.

CHAPTER 2

Candidate Structures for Inorganic Lithium Solid-state Electrolytes Identified by High-throughput Bond-valence Calculations

Ruijuan Xiao, Hong Li and Liquan Chen

Beijing National Laboratory for Condensed Matter Physics, Institute of Physics, Chinese Academy of Sciences, Beijing 100190, China

ABSTRACT

Looking for fast lithium ion conductors as solid state electrolytes is of great significance to achieve better safety for next generation lithium batteries. As an important prerequisite for developing all-solid-state lithium secondary batteries, the materials with high lithium ionic conductivity and inhibited electronic conductivity must be found. By implementing the bond-valence code and the automation simulation flow, we perform the high-throughput bond-valence calculations to screen fast lithium ion conductors from more than 1000 lithium-contained compounds in ICSD database. The candidate structures are identified and their kinetic properties as well as electronic structures are analyzed through bond-valence method and density functional theory calculations, respectively. The promising structures are selected to be further optimized in the future.

INTRODUCTION

Lithium rechargeable batteries are vital technology in today's modern society, and they have become widespread day by day in portable electronic devices [1], hybrid and electric vehicles [2], and large-scale energy storage systems for wind or solar energy [3].Looking for solid state electrolytes with high lithium ionic conductivity is an important precondition for developing all-solid-state lithium secondary batteries [4]. Because of the stability and non-flammability of inorganic solid electrolytes, the all-solid-state batteries are expected to exhibit less side reactions and higher safety, which are promising solutions for problems of leakage, vaporization, decomposition and safety of currently used lithium ionic batteries containing liquid electrolytes [5]. One of the most important technology for solid-state batteries is the discovery of solid electrolytes with good comprehensive performance, including high total ionic conductivity, wide electrochemical window, stable electrode/electrolyte interface, inhibited electronic conductivity, etc. [6] Among all the conditions, the high lithium ionic conductivity and the high electrical resistance are essential prerequisites because the former reduces the internal resistance of the battery and the later minimizes the self-discharge rate of the system. The experimental search of new lithium ionic conductor has been ongoing nearly half century. The fast lithium ion conduction is mainly found in glassy- and crystalline-sulfides [7] and [8], NASICON-, perovskite-, garnet- and γ-Li_3PO_4 type compounds [9], [10], [11] and [12], as well as Li_3N- and LiI-related structures [13] and [14]. To date, the highest lithium ionic conductivity in solid materials at room temperature is up to 12 mS/cm, exhibited in $Li_{10}GeP_2S_{12}$ and comparative to the value of liquid organic electrolytes [15], however, the air instability of sulfides limits the future application of this material. The experimental search involves numerous combinations of elements, compositions and crystal structures make the discovery of new solid-state electrolytes an arduous task.

Nowadays, as the development of efficient theoretical methods and inexpensive computers, the high-throughput theoretical calculations have played an increasingly important role in the discovery of new materials [16]. For the requirements needed by solid-state electrolytes, the electrical resistance can be estimated by the electronic structure calculations [17], and the inhibited electronic conductivity is expected to appear in wide bandgap compounds. While the simulation of migration process of lithium ions is much difficult since it is a kinetic process. The lithium ionic conductivity is closely related with the diffusion energy barrier which measures the minimum energy that must be input to Li^+ to complete a hop from one site to another in the structure. The calculations of energy barriers with high accuracy can be obtained from the transition-state method, like the nudged elastic band (NEB) method, or the molecular dynamic method based on density function theory (DFT) [17], [18] and [19]. These methods are reliable, and besides the energy barrier, they are able to reveal the diffusion mechanism and dominant type of carriers [20]. The disadvantages pointed out are the huge computational cost, which limit the integration of these methods into high-throughput simulation process. Thus, very few reports have involved the high-throughput density functional theory calculations in the screening or optimization of solid state electrolytes. So far the only work is carried out by Fujimura *et al.*, who carried out systematic sets of first-principles molecular dynamics simulations on LISICON-type electrolytes to optimize the ionic conductivities in this structure [21]. An alternative scheme is to adopt the simulation techniques in different levels of accuracy at various screening stages according to the distinctive features of each method. For example, the fast bond-valence (BV) technique is suitable for high-throughput pre-screening a wide range of compounds since the trend in the ability of ion motion can be drawn from the relative values of the migration energy barriers despite of their less accuracy compared with quantum mechanical simulations [22]. While the time-consuming DFT method can be adopted to do more precise calculations only for those promising candidates assigned by the BV method. For the derivative structures achieved by substitution or doping the existing compounds, the DFT computation is a powerful tool to predict exact structures which are important information for performing BV calculations. By combining the BV method and the DFT calculations, we proposed a high-throughput screening and optimization scheme to search fast lithium ion conductors as candidate solid state electrolytes [23]. In this work, the candidate structures for fast lithium ion conductors screened from the inorganic crystal structure database (ICSD) [24] by high-throughput bond-valence calculations are introduced and the electronic insulating properties are evaluated for some typical structures by DFT band structure calculations.

COMPUTATIONAL DETAILS

To perform the bond-valence calculations, we implement the computer code, BVpath, to calculate the migration energy barriers and pathways of lithium ions in inorganic crystals using the BV-based force-field approach proposed by Adams *et al.* [25]. In this method, the lithium ions are assumed to move within the framework composed of immobile anions and cations. Thus, the interactions between a dummy lithium ion in the space and the ions around it are empirically evaluated by the summation of the attraction between Li^+ and anions descried by Morse-type potential and the repulsion between Li^+ and cations expressed by Coulombic potential as follows [25],

$$E(Li)_{Morse} = D_0 \{ (\exp[\alpha(R_{min} - R)] - 1)^2 - 1 \} \tag{1}$$

$$E(Li - A)_{Coulomb} = \frac{q_{Li} q_A}{R_{Li-A}} \, erfc \left(\frac{R_{Li-A}}{\rho_{Li-A}} \right) \tag{2}$$

in which, D_0, α, R_{min} are Morse potential parameters determined from a large amount of stable compounds, q and R refers to the charge and Li-A distance, respectively. The summation of above two terms on grids of points with resolution of 0.1 Å builds the maps of the total potential energy, $E(Li)$, in a unit cell. Other kinds of empirical potential energy functions may selected for this calculations, while only Morse-type and Coulombic potentials are considered here because their extensive use have confirmed their effectiveness in a lot of systems [25]. The regions enclosed by isosurfaces of constant $E(Li)$ are considered as the space where lithium ions can go through, and the threshold value of $E(Li)$ to form a continuous pathway is estimated as the Li^+ migration energy barrier. In the calculations, the cutoff radius of 10 Å is taken for the interaction models. Although this method is limited in revealing the diffusion mechanisms in materials, it is effective to be used in the initial screening because it takes short computational time and it is easy to be extended to broad range of materials.

The BVpath code is written by Fortran and the calculation time for a single structure is a few minutes. To carry out the high-throughput bond-valence calculations for screening of a large structure database, we develop an automatic flow consisting of a set of scripts written in the Python programming language. The primary role of these scripts is to initiate the calculations, including generating the input files from the structures in the database, specifying the necessary parameters, submitting the simulation tasks, calling the BVpath code, and detecting the errors and returning the message with minimal human intervention during the whole process. With this high-throughput calculations based on BV methods, more than 1000 lithium-contained oxides from ICSD database are studied and their migration energy barriers and pathways are obtained. The effectiveness of the high-throughput bond-valence calculations and the ability to predict reliable tendency of the Li^+ migration energy barriers have been confirmed in Ref. [23] by comparing with the results from DFT calculations.

For the compounds with low migration energy barriers screened from bond-valence calculations, the band structures are simulated by density functional theory with the Vienna *ab initio* simulation package (VASP) [26] using the generalized gradient approximation (GGA) with a parameterized exchange-correlation functional according to PBE [27]. In the first step, the structural relaxation is performed with the k-mesh in the density of one point per 0.06 $Å^{-3}$. The cutoff for the wave function is specified as 30% larger than the maximal cutoff value among all elements involved in the compound. Based on the relaxed geometries, the band structures are calculated along the high-symmetry k-path in the Brillouin zones standardized by Setyawan and Curtarolo according to the Bravais lattices [28].

RESULTS AND DISCUSSION

Table 1 lists the lithium migration energy barriers, E_a, and the dimension of pathways for part of the oxides obtained by the high-throughput bond-valence calculations. The compounds listed from line 1 to 10 show the lowest energy barriers among all the calculated structures, and the lithium migration pathways of those with E_a less than 0.5 eV are illustrated in Figure 1. The one dimensional channel with very low E_a value of 0.25 eV is found in the orthorhombic LiB_3O_5, which was considered as a nonlinear optical crystal [29] and the pyroelectric, dielectric, and piezoelectric properties of this material have been widely investigated [30]. It is hardly to find any reports about the ionic conductivity of this compound except the work from Radaev *et al.*, in which the atomic structure is identified by X-ray diffraction and the one-dimensional ionic conductivity is deduced [31]. As shown in Figure 1(a), the lithium ions can diffuse along c axis within the space around by B-O network. The electronic structure of LiB_3O_5 has been computed from first principles method and a direct bandgap of 7.37 eV is characterized by Xu and Ching [32]. The wide bandgap is expected to ensure the broad electrochemical window of this material. Thus, it is worth to further study the lithium ionic conductivity of this material in the future. The rhombohedral Li_2O also shows low E_a value of 0.25 eV, and it is one of the component units in the fast lithium ion conducting glass [33], and may also act as the buffer layer between solid electrolyte and electrodes. One report on ionic conductivity has been found for α-$Li_2Te_2O_5$ with monoclinic lattice and β-$Li_2Te_2O_5$ with orthorhombic structure (not listed in the table because of the larger E_a value), and the former shows higher ionic conductivity than the later one [34]. The experimentally measured energy barrier of α-$Li_2Te_2O_5$ reported in Ref. [34] is higher than the prediction by BV method, and no other experimental or theoretical reports can be found for comparison, therefore, more detailed work is necessary to understand the kinetic properties of this compound. LiB $(C_2O_4)_2$, abbreviated as LiBOB, is an electrolyte salt already used in the lithium ion batteries [35]. Some works are focused on the diffusion properties of lithium-silicate glass [36], and a theoretical simulation has been performed on crystalline $Li_2Si_2O_5$ [37]. Li_2SO_4 has been confirmed as both a lithium-ion conductor and a proton conductor, and used as aqueous electrolyte in lithium ion batteries [38]. Both the glassy and crystalline state of $LiAlSiO_4$ have been studied experimentally and the glassy state shows a little lower energy barrier than that in crystalline phase [39]. β-Li_5AlO_4 in orthorhombic lattice is known as humidity sensor and CO_2 captor [40] and [41], thus the application of this material in solid state batteries may be limited because of the complex reactions between the air and itself. However, the ionic conductivity up to 10^{-3} S/cm at 300°C in substituted Li_5MO_4 phase (M=Al, Ga, Fe) can be obtained [42] and [43], which provide opportunities to optimize the material. Among the isomer of Li_5AlO_4, α-phase is more stable in low-temperature but shows higher E_a of 0.75 eV in three-dimensional paths, which is also a candidate structure for fast lithium ion conductors if suitable optimization scheme can be found. We carry out the DFT calculations for α-Li_5AlO_4 and β-Li_5AlO_4, and the band structures are given in Figure 2. Both of these two structures belong to orthorhombic lattice, thus the same high-symmetry k-path is selected for them in the band structure calculations. The calculated bandgap is 4.65 eV and 4.79 eV for α- and β-Li_5AlO_4, respectively, providing guarantee for inhibited electronic conductivity and moderate potential range. Therefore, the orthorhombic Li_5MO_4 is worth to be thoroughly studied to optimize the overall performance of this structure. $Li_6Si_2O_7$ is another lithium silicate may exhibit good kinetic properties. It has been studied in the glassy state of $Li_4B_2O_5$-$Li_6Si_2O_7$-$Li_4P_2O_7$ system [44], but there is no report on the ionic conductivity of its crystalline phase. Besides the compounds with E_a values less than 0.5 eV, some others structures may behave as good ionic conductors are also listed in Table 1. It has been noticed that in most cases the E_a value by BV method is always larger than that calculated by DFT calculations because the relaxation of the structural frame during the lithium migration is ignored in the BV-based simulations [23]. Accordingly, we choose a relatively large value of E_a, up to 1.0 eV, as the threshold for screening fast lithium ion conductors with BV method. Several typical candidate structures will be discussed in the next subsections.

Table 1. Some candidate structures for fast lithium ion conductors screened from ICSD database by high-throughput bond-valence calculations.

No.	Compound name	Lattice system	Space group	E_a by BV [eV]	Path-way	ICSD code
1	LiB_3O_5	Orthorhombic	$P\,n\,a\,2_1$	0.25	1D	415199
2	Li_2O	Rhombohedral	$R\,\text{-}3\,m$:h	0.25	3D	108886
3	$Li_2Te_2O_5$	Monoclinic	$P1\,2_1/n\,1$	0.39	1D	26451
4	$LiB(C_2O_4)_2$	Orthorhombic	$P\,n\,m\,a$	0.4	1D	281623
5	$Li_2Si_2O_5$	Orthorhombic	$P\,b\,c\,n$	0.47	1D	67110
6	Li_2SO_4	Monoclinic	$P1\,2_1/c\,1$	0.48	2D	2512
7	$LiAlSiO_4$	Hexagonal	$P\,6_2\,2\,2$	0.49	1D	32595
8	Li_5AlO_4	Orthorhombic	$P\,m\,m\,n$	0.49	2D	1037
9	$Li_6Si_2O_7$	Tetragonal	$P\,\text{-}4\,2_1m$	0.49	3D	25752
10	Li_4CO_4	Monoclinic	$C\,1\,m\,1$	0.5	2D	245392
11	$LiScP_2O_7$	Monoclinic	$P\,1\,2_1\,1$	0.55	1D	91496
12	$LiInP_2O_7$	Monoclinic	$P\,1\,2_1\,1$	0.73	1D	60935
13	Li_2CaGeO_4	Tetragonal	$I\,\text{-}4\,2\,m$	0.58	1D	19024
14	Li_2CaSiO_4	Tetragonal	$I\,\text{-}4\,2\,m$	0.59	1D	19023
15	Li_2ZnSiO_4	Monoclinic	$P\,1\,2_1/n\,1$	0.9	3D	8237
16	$LiScGeO_4$	Orthorhombic	$P\,n\,m\,a$	0.85	1D	62481
17	$LiInGeO_4$	Orthorhombic	$P\,n\,m\,a$	0.87	1D	62229
18	$LiZnPO_4$	Orthorhombic	$P\,n\,2_1a$	0.87	1D	79352
19	$LiMgPO_4$	Orthorhombic	$P\,n\,m\,a$	0.88	1D	201138
20	$LiInSiO_4$	Orthorhombic	$P\,b\,n\,m$	0.97	1D	281318
21	$LiScSiO_4$	Orthorhombic	$P\,b\,n\,m$	0.99	1D	77544
22	Li_2GeO_3	Orthorhombic	$C\,m\,c\,2_1$	0.55	1D	100403
23	Li_2SiO_3	Orthorhombic	$C\,m\,c\,2_1$	0.58	1D	100402
24	Li_2SnO_3	Monoclinic	$C\,1\,2/c\,1$	0.76	3D	21032
25	$Li_3Sc_2(PO_4)_3$	Monoclinic	$P\,1\,1\,2_1/n$	0.76	2D	86457
26	$Li_3In_2(PO_4)_3$	Monoclinic	$P\,1\,2_1/n\,1$	0.83	2D	60948
27	$LiGe_2(PO_4)_3$	Triclinic	$R\,\text{-}3\,c$:h	0.88	3D	69763
28	$LiZr_2(PO_4)_3$	Monoclinic	$P\,1\,2_1/n\,1$	0.89	2D	91112

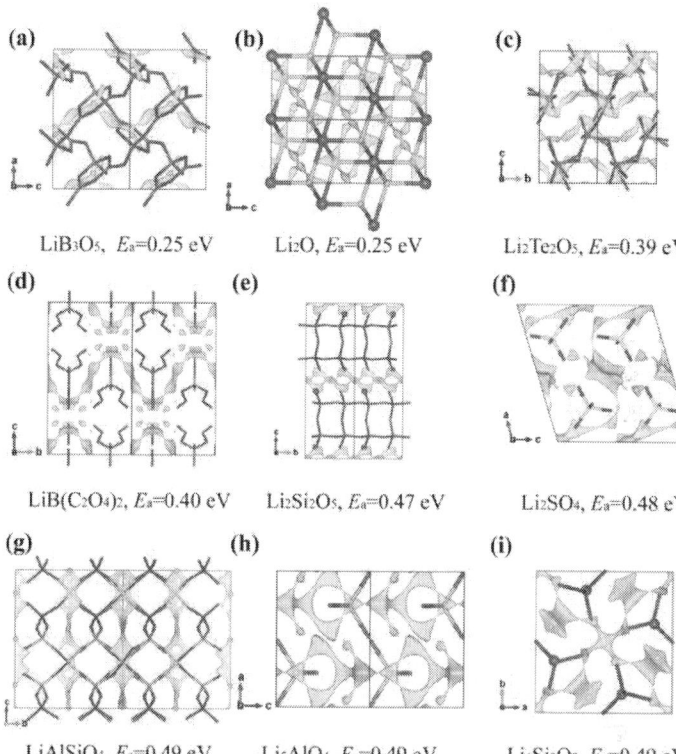

Figure 1. The lithium migration pathways reveal by BV calculations for the candidate structures with migration energy barriers E_a less than 0.5 eV.

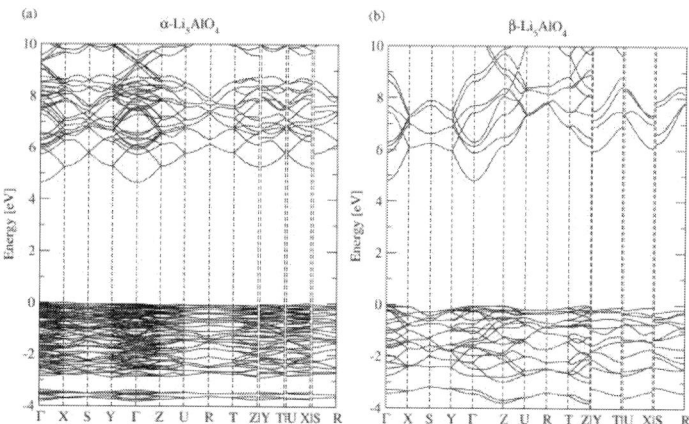

Figure 2. The band structure calculated by DFT method for (a) α-Li₅AlO₄ and (b) β-Li₅AlO₄. The Fermi energy is referred to zero and the high-symmetry k-path is selected according to Ref. [28].

Li4CO4

Carbonates and orthocarbonates are always interesting objects in chemistry. Li_2CO_3 is an ionic conductor with "knock-off" diffusion mechanism and acts as the main inorganic component of solid electrolyte interphase [45]. Orthocarbonates, building with $[CO_4]^{4-}$ units, can be considered to be formed by the double hydration of CO_2, have been computationally demonstrated stable in S_4 symmetry [46], although still not been observed experimentally. The monoclinic-Li_4CO_4 presented here is predicted to exist at high pressure [47]. According to the E_a value calculated by BV method, the energy needed to overcome for Li^+ migrating in monoclinic-Li_4CO_4 is 0.50 eV. As shown inFigure 3, the $[CO_4]^{4-}$ units are arranged in ab plane with the same orientation in this structure. Li^+ ions locate within the $[CO_4]^{4-}$ layer or between the neighboring layers. The former is difficult to migrate because of the blocking from $[CO_4]^{4-}$ units around Li^+, while the latter is easy to move along the two-dimensional paths in the lithium layer. The electronic structure calculation indicates that the bandgap is about 4 eV, and the highest valence band is mainly composed of O-p states, illustrated by the density of states shown in Figure 4. The hybridization between C and O is obvious at -6 eV and -8 eV, but the interactions between Li and O (or C) is relatively weak. The distortion of the structure and the charge transfer between Li^+ and $[CO_4]^{4-}$ units during Li^+ movement is needed to be clarified to further understand the stability of the compound and figure out a way to synthesize the structure. There are some other Li_4CO_4 isomers also shown low migration energy barriers, including tetragonal structure and several monoclinic lattices with different $[CO_4]^{4-}$ orientations. These structures should be further studied and may be used as the start points for the optimization of properties.

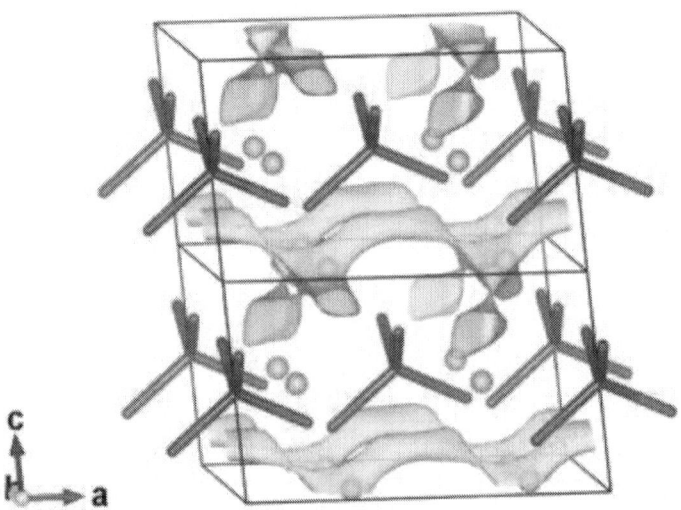

Figure 3. The two-dimensional migration pathways in monoclinic-Li_4CO_4 calculated by BV method.

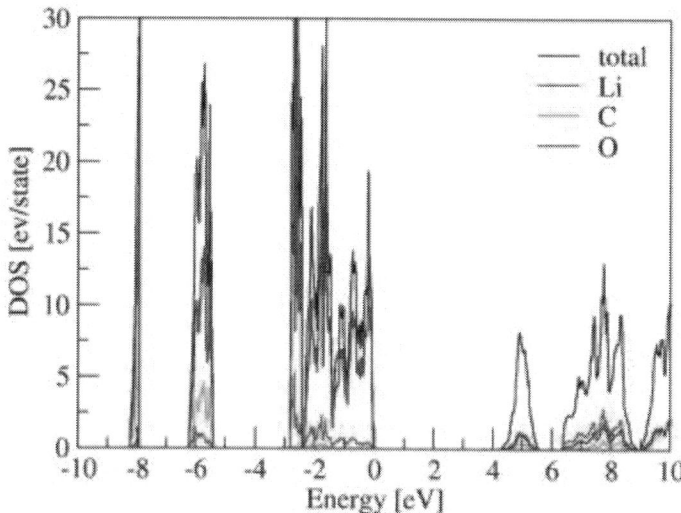

Figure 4. The total and partial density of states for monoclinic-Li$_4$CO$_4$. The Fermi energy is referred to zero.

Li$_2$MXO$_4$

The γ-tetrahedral structure, Li$_2$MXO$_4$ (M=Zn, Mg, Ca and X= Ge, Si, Ti), is isostructural with γ-Li$_3$PO$_4$, which already attracts a lot of attention as a solid-state electrolyte material for batteries [48]. In this structure, the alternating MO$_8$ dodecahedra and XO$_4$ tetrahedra are sharing edges and linked to columns. Neighboring columns are connected at corners to form a three-dimensional network. Lithium ions are enclosed in the channels parallel to c axis, which provides a one-dimensional pathway for Li$^+$ migration. The E_a value calculated by BV method is 0.58 and 0.59 eV for Li$_2$CaGeO$_4$ and Li$_2$CaSiO$_4$, respectively, as indicated in line 13 and 14 in Table 1. Previous experiments revealed that tetrahedral-Li$_2$ZnSiO$_4$ shows dc conductivity of 10^{-6} S/cm at room temperature [49], implying that there is still much room for improvement. Figure 5 is the band structure for Li$_2$CaSiO$_4$ in tetrahedral lattice. The energy gap is larger than 5 eV, evincing the electrochemical stability of this material. As a promising candidate structure for solid state electrolytes, it is necessary to analyze all possible doping strategies in detail for improving the lithium ionic conductivities.

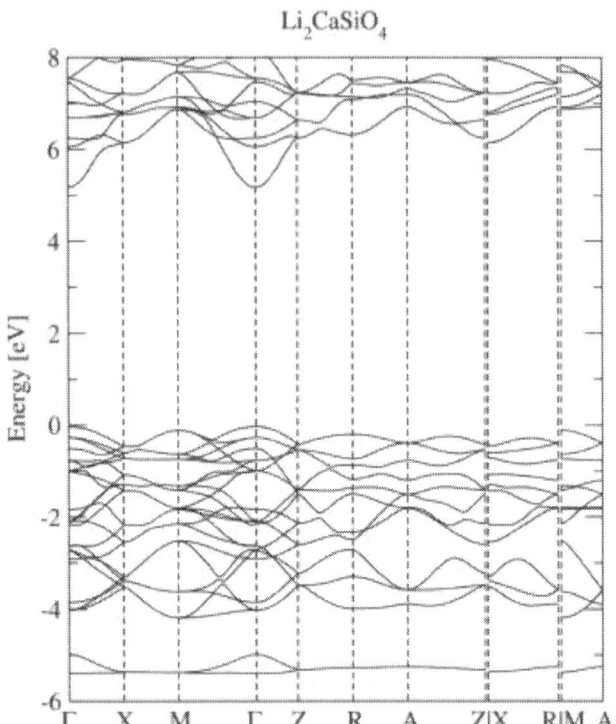

Figure 5. The band structure of Li₂CaSiO₄ calculated by DFT method. The Fermi level is referred to 0 eV and the high-symmetry k-path is chosen according to Ref. [28].

LiMXO₄, Li₂MO₃ and LiM₂(XO₄)₃

Besides increasing the lithium ionic conductivity of electrolytes, the reduction of interfacial resistance between electrolyte and electrodes is another challenge for designing the solid-state batteries. One hopeful scheme to improve the interfacial conduction is to construct the interface by two crystals with similar structures and orientation. According to this idea, we find some candidate structures to pair with those attractive cathode materials, including LiMXO₄, Li₂MO₃ and LiM (PO₄)₃ as listed in Table 1 from line 16 to 28.

Olivine-type structure LiMXO₄ shows same structure with LiFePO₄, which is a commercialized cathode with high safety, environmental benignity and low cost. The BV-method calculated E_a values for LiScGeO₄, LiInGeO₄, LiZnPO₄, LiMgPO₄, LiInSiO₄ and LiScSiO₄ are about 0.85~1.00 eV. The migration of lithium ions is along the one-dimensional path in b axis, shown in Figure 6(a), as what found in LiFePO$_4$ [50]. Jalem *et al.* adopted *ab initio* method to study 66 olivine-type oxides and identified Mg-As, Sc-Ge, In-Ge, Mg-P as promising M-X pairs to improve the ionic conductivity in this structure[51].

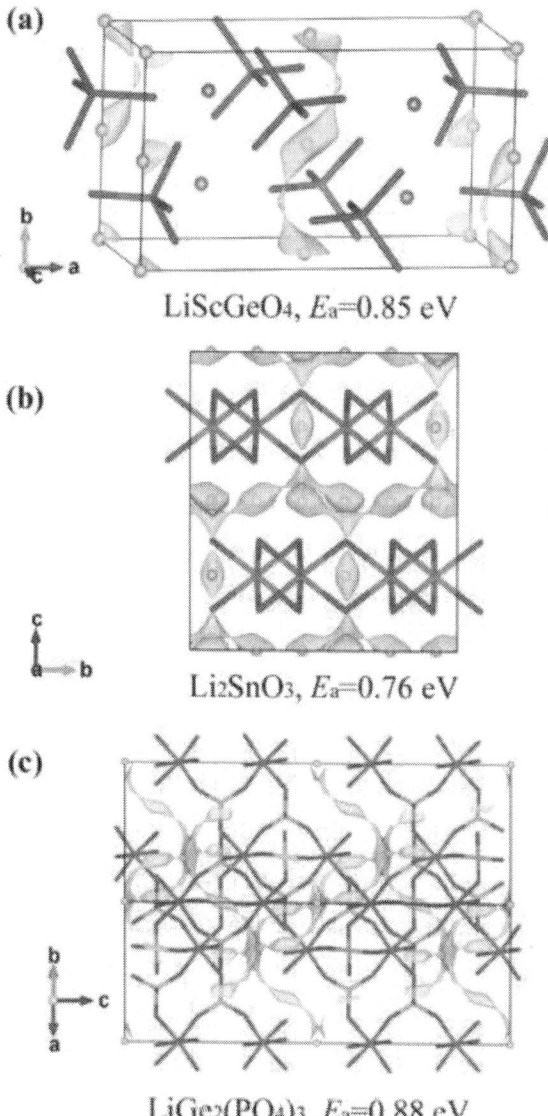

(a) LiScGeO$_4$, E_a=0.85 eV

(b) Li$_2$SnO$_3$, E_a=0.76 eV

(c) LiGe$_2$(PO$_4$)$_3$, E_a=0.88 eV

Figure 6. The lithium migration energy barriers and pathways calculated by BV method for (a) LiScGeO$_4$, (b) Li$_2$SnO$_3$ and (c) LiGe$_2$(PO$_4$)$_3$.

Li_2MnO_3 plays a key role in Li-rich Mn-based layered materials ($mLi_2MnO_3 \cdot nLiMO_2$, M = Mn, Ni, Co, etc.) for achieving unusually high lithium storage capacity [52]. DFT calculations revealed that the Li^+ hopping can happen within the Li plane and between Li and transition-metal layers with activation energies in the range from 0.51 eV to 0.84 eV[53]. By screening with high-throughput bond-valence calculations, Li_2SnO_3 with same structure and similar diffusion path to Li_2MnO_3 is detected as shown in Figure 6(b). The E_a value of Li_2SnO_3 calculated by BV method is 0.76 eV, and the lattice mismatch with Li_2MnO_3 is ~7%. Therefore, it is a promising candidate to pair with Li-rich cathode in solid-state batteries. The lattice constants of Li_2GeO_3 and Li_2SiO_3 are also close to those of Li_2MnO_3. However, the variation from MnO6 and SnO6 octahedra to GeO$_4$ and SiO$_4$tetrahedra changes the arrangements of the blocking units. As a result, the one-dimensional migration paths occur in Li_2GeO_3 and Li_2SiO_3.

$LiM_2(XO_4)_3$ in triclinic lattice is NASICON-type phase and already studied as solid electrolytes in previous investigations [54]. The solid electrolytes based on lithium titanium phosphate, *e.g.* $Li_{1.5}Al_{0.5}Ti_{1.5}(PO_4)_3$, have been applied in experiments to construct all-solid-state lithium batteries [55]. Our calculations indicates that the optimization of the NASICON-type $LiM_2(XO_4)_3$ may lead to new materials with high lithium ionic conductivity.

SUMMARY

We report the screening results of high-throughput bond-valence calculations and some candidate structures of fast lithium ion conductors are identified. By implementing the bond-valence simulation code, BVpath, and the automatic flow for high-throughput calculations, more than 1000 compounds from the ICSD database are studied. The promising compounds with low migration energy barriers are selected as original candidate structures for further optimization. In the candidates, LiB_3O_5 and $Li_2Te_2O_5$ are investigated little on the ionic conductivity previously, and worth to be studied in detail. Li_5AlO_4 and some lithium silicates have already noticed because of their high ionic conductivity. The thorough studies to optimize the overall performance of these structures are necessary. The orthocarbonates have been predicted to exist at high pressure and our calculations indicate the possible good ionic conductivity in these meta-structures. The efforts to synthesize and stabilize these structures maybe carried out in the future. The γ-tetrahedral structure Li_2MXO_4 also show very low migration energy barriers. Both the barrier and carrier concentration can be adjusted through substitution of M and X sites with atoms in different valence states and ionic radius. Several structures, which may well match the interface between electrolytes and electrodes, are also identified, including $LiMXO_4$, Li_2MO_3 and $LiM_2(XO_4)_3$ which show same structure with typical cathodes $LiFePO_4$, Li_2MnO_3 and $LiTi_2(PO_4)_3$, respectively. The well-matched structure may reduce the interfacial stress and improve the conduction. The bandgaps of the candidate structures are studied using DFT calculations. However, it is well known that the GGA functionals underestimate energy band gaps for many materials, while the meta-GGA functionals yield band gaps with an accuracy comparable with hybrid functional but with much less computational cost. Thus, the finer calculations will be carried out to confirm the band gaps with better accuracy in further investigations, and we will focus on the optimization of kinetic properties for the candidate structures and their derivatives.

ACKNOWLEDGEMENT

We acknowledge the National Natural Science Foundation of China (Grant Nos. 11234013 and 51172274), "863" Project (Grant No. 2015AA034201), and Beijing S&T Project (Grant No. Z13111000340000) for financial support and the Shanghai Supercomputer Center for providing computing resources.

REFERENCES

1. J.M. Tarascon, M. Armand, Issues and challenges facing rechargeable lithium batteries. Nature 414 (2001) 359-367.

2. E. Karden, S. Ploumen, B. Fricke, T. Miller, K. Snyder, Energy storage devices for future hybrid electric vehicles. J. Power Sources 168 (2007) 2-11.
3. B. Dunn, H. Kamath, J. M. Tarascon, Electrical energy storage for the grid: a battery of choices, Science 334 (2011) 928-935.
4. K. Takada, Progress and prospective of solid-state lithium batteries. Acta Mater. 6 (2013) 759-770.
5. K. Xu, Nonaqueous liquid electrolytes for lithium-based rechargeable batteries, Chem. Rev. 104 (2004) 4303-4417.
6. J. Gao, G. Chu, M. He, S. Zhang, R. J. Xiao, H. Li, L. Q. Chen, Screening possible solid electrolytes by calculating the conduction pathways using bond valence method, Sci. China-Phys. Mech. Astron. 57 (2014)1526-1535.
7. M. Ribes, B. Barrau, J.L. Souquet, Sulfide glasses: glass forming region, structure and ionic conduction of glasses in Na2S-XS2 (X=Si; Ge), Na2S-P2S5 and Li2S-GeS2 systems, J. Non-Crystal. Solids 38 (1980) 271-276.
8. R. Kanno, T. Hata, Y. Kawamoto, M. Irie, Synthesis of a new lithium ionic conductor, thio-LISICON–lithium germanium sulfide system, Solid State Ionics 130 (2000) 97-104.
9. J.-M. Winand, A. Rulmont, P. Tarte, Nouvelles solutions solides L I (M IV)2−x(N IV)x(PO4)3 (L = Li,Na M,N = Ge,Sn,Ti,Zr,Hf) synthèse et étude par diffraction x et conductivité ionique, J Solid State Chem. 93 (1991) 341-349.
10. A.G. Belous, G.N. Novitskaya, S.V. Polyanetskaya, Y.I. Gornikov, Study of complex oxides with the composition La2/3-xLi3xTiO3, Inorg. Mater. 23 (1987) 412.
11. V. Thangadurai, H. Kaak, W. Weppner, Novel fast lithium ion conduction in garnet-type Li5La3M2O12 (M= Nb, Ta), J Am. Ceram. Soc. 86 (2003) 437-440.
12. K. Kanehori, K. Matsumoto, K. Miyauchi, T. Kudo, Thin film solid electrolyte and its application to secondary lithium cell, Solid State Ionics 9 (1983) 1445-1448.
13. U. v. Alpen, A. Rabenau, G.H. Talat, Ionic conductivity in Li3N single crystals, Appl. Phys. Lett. 30 (1977) 621-622.
14. A.M. Stoneham, E. Wade, J.A. Kilner, A model for the fast ionic diffusion in alumina-doped LiI, Mater. Res. Bull. 14 (2012) 661-666.
15. N. Kamaya, K. Homma, Y. Yamakawa, M. Hirayama, R. Kanno, M. Yonemura, T. Kamiyama, Y. Kato, S. Hama, K. Kawamoto, A. Mitsui, A lithium superionic conductor, Nature Mater. 10 (2011) 682-686.
16. S. Curtarolo, G.L.W. Hart, M.B. Nardelli, N. Mingo, S. Sanvito, O. Levy, The high-throughput highway to computational materials design, Nature Mater. 12 (2013) 191-201.
17. J. Hafner, C. Wolverton, G. Ceder, Toward computational materials design: the impact of density functional theory on materials research. MRS Bull. 31 (2006) 659-668.
18. Y. S. Meng, M.E. Arroyo-de Dompablo, First principles computational materials design for energy storage materials in lithium ion batteries, Energy Environ. Sci. 2 (2009) 589-609.

19. C.Y. Ouyang, L.Q. Chen, Physics towards next generation Li secondary batteries materials: A short review from computational materials design perspective, Sci. China-Phys. Mech. Astron. 56 (2013) 2278-2292.
20. S.Q. Shi, P. Lu, Z.Y. Liu, et al. Direct calculation of Li-Ion transport in the solid electrolyte interphase. J Am. Chem. Soc. 134 (2012) 15476-15487.
21. K. Fujimura, A. Seko, Y. Koyama, A. Kuwabara, I. Kishida, K. Shitara, C.A.J. Fisher, H. Moriwake, I. Tanaka, Accelerated materials design of lithium superionic conductors based on first-principles calculations and machine learning algorithms, Adv. Energy Mater. 3 (2013) 980-985.
22. S. Adams, R.P. Rao, High power lithium ion battery materials by computational design, Phys. Status Solidi A 208 (2011) 1746-1753.
23. R.J. Xiao, H. Li, L.Q. Chen, High-throughput design and optimization of fast lithium ion conductors by the combination of bond-valence method and density functional theory (submitted) Inorganic Crystal Structure Database, ICSD. Karlsruhe: Fachinformationszentrum, 2008.
24. S. Adams, Relationship between bond valence and bond softness of alkali halides and chalcogenides, Acta Cryst. B57 (2001) 278-287.
25. G. Kresse, J. Furthmüller, Efficiency of ab-initio total energy calculations for metals and semiconductors using a plane-wave basis set, Comput. Mater. Sci. 6 (1996) 15-50.
26. J.P. Perdew, K. Burke, M. Ernzerhof, Generalized gradient approximation made simple, Phys. Rev. Lett. 78 (1997) 1396.
27. W. Setyawan, S. Curtarolo, High-throughput electronic band structure calculations: challenges and tools, Comput. Mater. Sci. 49 (2010) 299-312.
28. S. Lin, Z. Sun, B. Wu, C. Chen, The nonlinear optical characteristics of a LiB3O5 crystal, J. App. Phys. 67 (1990) 634-638.
29. R. Guo, S. A. Markgraf, Y. Furukawa, M. Sato, A. S. Bhalla, Pyroelectric, dielectric, and piezoelectric properties of LiB3O5, J. App. Phys. 78 (1995) 7234-7239.
30. S.F. Radaev, N.I. Sorokin, V.I. Simonov, Atomic structure and one-dimensional ionic conductivity of lithium triborate LiB3O5, Soviet physics. Solid state 33 (1991) 3597-3600.
31. Y. N. Xu and W. Y. Ching, Electronic structure and optical properties of LiB3O5, Phys. Rev. B 41 (1990) 5471-5474.
32. J. Fu, Fast Li+ ion conduction in Li2O-Al2O3-TiO2-SiO2-P2O5 glass-ceramics, J. Am. Ceram. Soc., 80 (1997) 1901-1903.
33. D. Cachau-Herreillat, A. Norbert, M. Maurin, E. Philippot, Comparative crystal chemical study and ionic conductivity of two varieties of α- and β-lithium tellurite (Li2Te2O5), J. Solid State Chem. 37 (1981) 352-361.
34. K. Xu, S. S. Zhang, U. Lee, J. L. Allen, T. R. Jow, LiBOB: is it an alternative salt for lithium ion chemistry, J. Power Sources, 146 (2005) 79-85.
35. S. Sen, A. M. George, J. F. Stebbins, Ionic conduction and mixed cation effect in silicate glasses and liquids: 23Na and 7 Li NMR spin-lattice relaxation and a multiple-barrier model for percolation, J. Non-Crystalline Solids, 197 (1996) 53-64.

36. K. M. Kennedy, G. Morrison, A. V. Chadwick, A computer modelling study of ionic conductivity in some lithium silicates and aluminosilicates, Anales de quimica-international edition, 94 (1998) 27-30.
37. P. Knauth, Ionic conductor composites: theory and materials, J. Electroceramics 5 (2000) 111-125.
38. A. Pechenik, D. H. Whitmore, S. Susman, M. A. Ratner, Transport in glassy fast-ion conductors: a study of $LiAlSiO_4$ glass, J. Non-Crystalline Solids, 101 (1988) 54-64.
39. R. T. Johnson Jr, R. M. Biefeld, Ionic conductivity of Li_5AlO_4 and Li_5GaO_4 in moist air environments: potential humidity sensors, Materials Research Bulletin 14 (1979) 537-542.
40. T. Avalos-Rendon, J. Casa-Madrid, H. Pfeiffer, Thermochemical capture of carbon dioxide on lithium aluminates ($LiAlO_2$ and Li_5AlO_4): a new option for the CO_2 absorption, J. Phys. Chem. A 113 (2009), 6919-6923.
41. T. Esaka, M. Greenblatt, Lithium ion conduction in substituted Li_5MO_4, M=Al, Fe, J. Solid State Chem. 71 (1987) 164-171.
42. T. Esaka, M. Greenblatt, Lithium ion conduction in substituted Li_5GaO_4, Solid State Ionics 21 (1986) 255-261.
43. Y. Kowada, M. Tatsumisago, T. Minami, Raman-spectra of rapidly quenched glasses in the systems Li_3BO_3-Li_4SiO_4-Li_3PO_4 and $Li_4B_2O_5$-$Li_6Si_2O_7$-$Li_4P_2O_7$, J. Phys. Chem. 93 (1989) 2147-2151.
44. Shi S Q, Qi Y, Li H, et al. Defect thermodynamics and diffusion mechanisms in Li_2CO_3 and implications for the solid electrolyte interphase in Li-ion batteries. J Phys. Chem. C, 2013, 117(17): 8579-8593.
45. N. Narasimhamurthy, H. Manohar, A.G. Samuelson, J. Chandrasekhar, Cumulative anomeric effect: a theoretical and x-ray diffraction study of orthocarbonates, J. Am. Chem. Soc. 112 (1990) 2937- 2941.
46. Ž.P. Čančarević, J.C. Schön, M. Jansen, Possible existence of alkali metal orthocarbonates at high pressure, Chem.- A Eur.J. 13 (2007) 7330-7348.
47. Y. A. Du, N. A. W. Holzwarth, Li ion diffusion mechanisms in the crystalline electrolyte γ-Li_3PO_4, J. Electrochem. Soc. 154 (2007) A999-1004. ACCEPTED ACCEPTED MANUSCRIPT 18
48. S. B. R.S. Adnan, N. S. Nohamed, Conductivity and dielectric studies of Li_2ZnSiO_4 ceramic electrolyte synthesized via citrate sol gel method, Int. J. Electrochem. Sci., 7 (2012) 9844-9858.
49. Y. Sun, X. Lu, R. J. Xiao, H. Li, X. J. Huang, Kinetically controlled lithium-staging in delithiated $LiFePO_4$ driven by the Fe center mediated interlayer Li-Li interactions, Chem. Mater. 2012, 24, 4693-4703.
50. R. Jalem, T. Aoyama, M. Nakayama, M. Nogami, Multivariate method-assisted ab initio study of olivine-type $LiMXO_4$(Main group M2+-X5+ and M3+-X4+) compositions as potential solid electrolytes, Chem. Mater. 24 (2012) 1357−1364.
51. M.M. Thackeray, C.S. Johnson, J.T. Vaughey, N. Li, S.A. Hackney, Advances in manganese-oxide 'composite' electrodes for lithium-ion batteries, J. Mater. Chem. 15 (2005) 2257-2267.
52. R. J. Xiao, H. Li, L. Q. Chen, Density functional investigation on Li_2MnO_3, Chem. Mater. 24 (2012) 4242-4251.

53. G. Nuspl, T. Takeuchi, A. Weiss, H. Kageyama, K. Yoshizawa, T. Yamabe, Lithium ion migration pathways in LiTi2(PO4)3, J. Appl. Phys. 86 (1999) 5484-5491.
54. G. Nuspl, T. Takeuchi, A. Weiss, H. Kageyama, K. Yoshizawa, T. Yamabe, Lithium ion migration pathways in LiTi2(PO4)3, J. Appl. Phys. 86 (1999) 5484-5491.
55. M. Kotobuki, Y. Isshiki, H. Munakata, K. Kanamura, All-solid-state lithium battery with a three-dimensionally ordered Li1.5Al0.5Ti1.5(PO4)3 electrode, Electrochimica Acta 55 (2010) 6892-6896.

CITATION

Ruijuan Xiao, Hong Li, Liquan Chen, Candidate structures for inorganic lithium solid-state electrolytes identified by high-throughput bond-valence calculations, Journal of Materiomics, Available online 22 August 2015, ISSN 2352-8478, http://dx.doi.org/10.1016/j.jmat.2015.08.001.

CHAPTER 3

Comparison of Photoluminescence Properties of GD2O3 Phosphor Synthesized by Combustion and Solid State Reaction Method

Raunak Kumar Tamrakar[1], Durga Prasad Bisen[2] and Nameeta Brahme[2]

[1]Department of Applied Physics, Bhilai Institute of Technology (Seth Balkrishan Memorial), Near Bhilai House, Durg (C.G.) 491001, India
[2]School of Studies in Physics and Astrophysics, Pt. Ravishankar Shukla University, Raipur (C.G.) 492010, India

ABSTRACT

In this paper, we presented a comparison of Photoluminescence (PL) studies of Gd_2O_3 phosphor prepared by two synthesis methods, high temperature (1400 °C) solid state reaction and low temperature combustion synthesis. Both of these methods synthesized Gd_2O_3 phosphor successfully. Structural properties of Gd_2O_3 phosphor were investigated by X-ray diffraction (XRD), Fourier transform infrared spectroscopy (FTIR). Scanning electron microscope (SEM), Energy dispersive X-ray analysis (EDX), and Transmission electron microscopy (TEM). The results obtained for both methods of preparation were compared. The photoluminescence (PL) properties of pure Gd_2O_3 were carried out over the range of 300 nm–650 nm. A PL emission band at UV region in between 318 and 370 nm along with blue, green and red emission was observed at 275 nm excitation.

INTRODUCTION

Over the past decade, Trivalent rare earth oxides (RE_2O_3) are used in high performance luminescence devices, magnet, catalysts and other functional materials because of their optical, chemical and electronic characteristics resulting from the 4f electronic shell as most stable compound. These phosphor materials having wide applications range from fluorescent lamp to luminescence devices (Guo et al., 2004a, Guo et al., 2004b, Guo et al., 2004c, Korzenski et al., 2001 and Dhananjaya et al., 2011). They essentially convert one type of energy into visible, UV and near infrered radiation, means they work as highly efficient optical transducer. They are generally crystalline with lots of defect and impurity in nature with low or efficient luminescence properties; we have a challenge to enhance their luminescence properties by applying different synthesis and calcination process. As compared to other oxide material, recently a variety of rare earth oxides (REO) have been the subject of intense due to its unique properties as well their potential application (Dosev et al., 2006, Pang et al., 2003 and Tamrakar et al., 2014c).

Among all rare earth oxide, especially gadolinium oxide has much attention widely investigated and attracted by the researcher because of several properties particularly physical properties, such as its crystallographic stability up to 2325 °C (melting point of Gd_2O_3 is greater than 2400 °C), high mechanical strength, high thermal conductivity ($0.1\ Wcm^{-1}K^{-1}$), large optical band gap (5.4 eV). As a luminescent material high dielectric constant ($k \approx 18$), a rather high refractive index near $n \approx 2$ which is well suited for waveguide applications, and also having outstanding optical, mechanical, and chemical properties as well as low phonon energy of $519.9\ cm^{-1}$ (Li et al., 2006, Lushchik et al., 2003, Tamrakar et al., 2014a, Tamrakar et al., 2014b, Tamrakar et al., 2014c, Tamrakar et al., 2014d and Wang et al., 2006). Therefore, Gd_2O_3 has been shown to be suitable for rare earth oxide materials that emit both in visible and UV regions and has received considerable attention for sensor, opto-electronics, data storage devices, FPD panel and luminescence application. Various methods, including hydrothermal, precipitation, combustion, sol-gel and spray pyrolysis, have been employed to fabricate Gd_2O_3phosphors (Tamrakar, 2012 and Tamrakar et al., 2014a). Among them, combustion is one of the attractive methods due to the advantages of simplicity, high efficiency, energy saving and uniform morphology. It is also possible to achieve a better control of stoichiometry and specific properties of the final product (Tamrakar, 2012, Tamrakar et al., 2014a and Zou et al., 2014). The solid state reaction method is economical, simple and more versatile than the others are. This method gives the possibility of obtaining phosphor with suitable properties for luminescent applications and when large areas are needed (Tamrakar, Bisen, Robinson, et al., 2014). Rare earth sesquoxides exist in different structural types depending on the ionic radii of the rare earth ion. It is having great interest is due to the unique properties of materials with dimensions less than 100 nm. One of the most fascinating and useful aspects is change in their optical properties, including linear and nonlinear absorption, photoluminescence, electroluminescence and light scattering as a function of size and synthesis methods. So shape, size, interaction between particles and synthesis technique can also play an important role for it optical behavior (Bhargave and Gallagher, 1994, Tamrakar, 2012, Tamrakar and Bisen, 2013a and Wang et al., 2006). Therefore, it is quite interesting to study the change in optical properties that occur in these oxides when subjected to high temperature (Pang et al., 2003, Tamrakar et al., 2014b and Tamrakar et al., 2014c; Tamrakar, Kanchan, & Bisen, 2014).

The objective of the present work is to compare the Gd_2O_3 phosphor prepared by both solid state reaction method and combustion synthesis. The luminescence behaviors of the prepared phosphors were compared for both the methods. The phase purity and the morphology of the samples prepared by both synthetic routes were compared.

EXPERIMENTAL

Solid State Reaction

Conventional solid state method was used for the synthesis of the phosphor. The prepared phosphor powder was utilized without any further treatment. High purity gadolinium nitrate Gd $(NO_3)_3$, (99.99%) purchased from sigma Aldrich and was used as starting material. The fixed amount of

Gd $(NO_3)_3$ was weighed and ground into a fine powder by agate mortar and pestle. The grinded sample was placed in an alumina crucible and heated at 1100 °C for 1 h followed by dry grinding and further heated at 1400 °C for 4 h in a muffle furnace. The sample was allowed to cool at room temperature in the same furnace for about 15 h (Tamrakar et al., 2014b, Tamrakar et al., 2014c and Tamrakar et al., 2014d).

Combustion Synthesis

For synthesis of Gd_2O_3 phosphor, an aqueous precursor solution containing Gd $(NO_3)_3 \cdot 6H_2O$ and urea were used. A constant percentage of urea and Gd $(NO_3)_3$ in ratio of 2:1 was mixed to prepare the precursor solution. When the solution was stirred for 1 h, it was transformed into a transparent gel. The solution was concentrated by heating until excess free water evaporated and spontaneous ignition occurred. In the beginning, solution undergoes dehydration with the liberation of large amount of gaseous products. At the point of spontaneous combustion, the solution begins burning and releases lot of heat; the solution vaporizes instantly and becomes a burning solid with liberation of gaseous by products such as oxide of carbons and nitrogen the combustion was finished. When it cooled down to room temperature, the resultant particles were crushed by mortar and pestle; then annealed at different temperatures 600 °C, 700 °C, 800 °C and 900 °C for 2 h (Sun et al., 2000, Tamrakar and Bisen, 2013b, Tamrakar et al., 2014a,Tamrakar et al., 2014b, Tamrakar et al., 2014c and Yanhong and Guangyan, 2007).

The crystallinity as well as the particle size of the phosphor was monitored X-ray diffraction measurement. X-ray diffraction (XRD) is a popular and powerful technique for determining crystal structure of crystalline materials. So Crystal structure and phase formation of the phosphors were examined using an X-ray diffractometer (Philips PAN Analytical X'pert Pro) operating at 40 kV and 30 mA with CuKα radiation ($\lambda = 1.54056$ Å). Molecular structure was determined by FTIR analysis done by Nicolet Instruments Corporation USA MAGNA-550. The surface morphology of the prepared phosphor was determined by field emission scanning electron microscopy (FESEM) JSM-7600F. Energy dispersive X-ray analysis (EDX) was used for elemental analysis of the phosphor. Transmission Electron Microscopy (TEM) using Philips CM-200 determined particle diameter and surface morphology of prepared phosphor. The PL emission spectra were recorded with a spectrofluorophotometer (SHIMADZU, RF-5301 PC), which was also used to record the excitation spectra.

RESULT AND DISCUSSION

X-ray Diffraction Analysis (XRD)

The X-ray powder diffraction (XRPD) analysis was used to characterize the synthesized phosphors. The resultant diffraction patterns for phosphors prepared by both solid state reaction method and combustion synthesis method revealed the presence of cubic Gd_2O_3 and monoclinic Gd_2O_3 respectively (Fig. 1A and B). The Bragg reflections for Gd_2O_3 by combustion method matched those of the JCPDS no. 43-1015(Grier & Mccarthy, 1991) and for Gd_2O_3 by solid state matched those of the JCPDS no 43-1014 (Grier & Mccarthy, 1991). The presence of cubic phase in solid state synthesis is most likely due to the fact that the temperature profile involved in the solid state method favors the production of cubic system. Furthermore, an analysis of the XRPD pattern was used to evaluate the crystalline size of the prepared Gd_2O_3 phosphors using Scherrer's equation (Guinier, 1963): The size of the particle has been computed from the full width half maximum (FWHM) of the intense peak using Debye Scherer formula.

$$D = \frac{k\lambda}{\beta \cos\theta}$$

Here, D is particle size

β is FWHM (full width half maximum)

$\lambda = 1.53$ Å is the wavelength of X-ray source (Cu (K_α) radiation)

θ is Bragg angle of the X-ray diffraction peak.

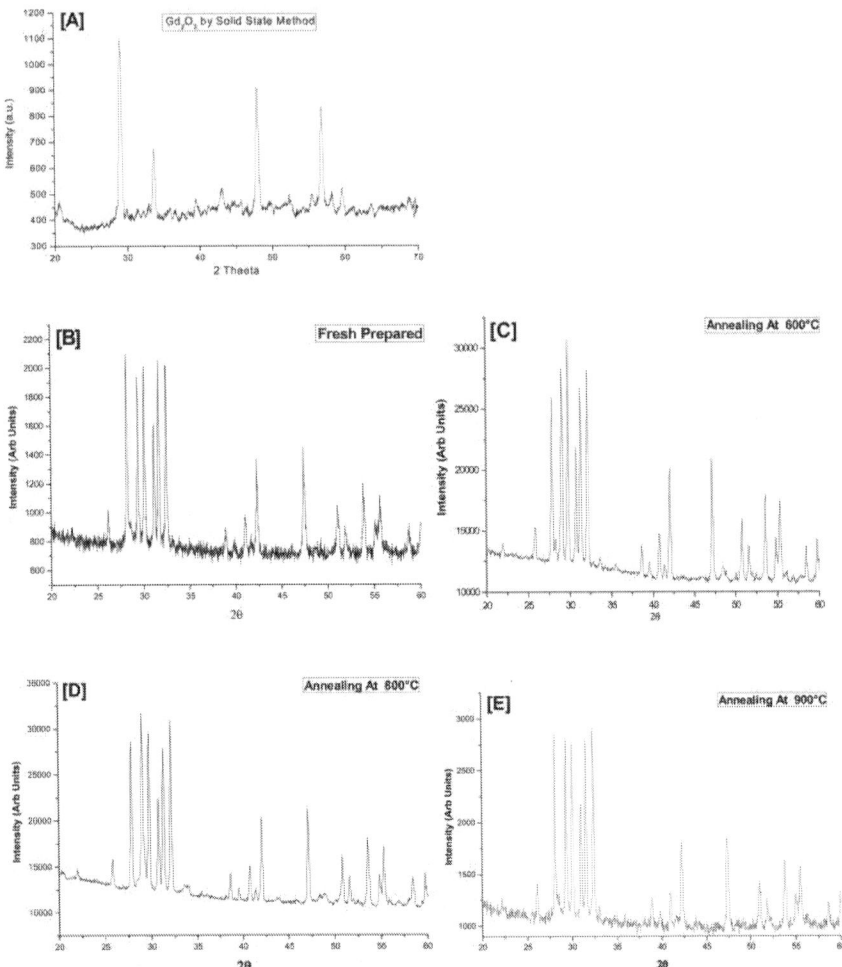

Figure 1. A) XRD patterns of pure Gd₂O₃ by solid state method. B–E) XRD patterns of pure Gd₂O₃ by combustion synthesis and canclinated at different temperatures 600 °C–900 °C.

For the Gd₂O₃ phosphor prepared by solid state method, the crystal size was found approximately 40 nm, while those prepared by combustion synthesis were in the range of 8 nm-30 nm. The particle size of the Gd₂O₃ phosphor increases with increasing annealing temperature (Tamrakar et al., 2014a, Tamrakar et al., 2014c and Tamrakar et al., 2014d).

Fourier Transform Infrared Spectroscopy (FTIR) Results

In order to determine the atomic bonds in a molecule FTIR analysis was carried out. FTIR spectra of Gd_2O_3 prepared by both methods have almost similar peaks in the range 400 cm^{-1} to 1000 cm^{-1}. The bands around 542 and 440 cm^{-1} are assigned to the Gd–O vibration of Gd_2O_3 (Garcia Murillo et al. 2001, Tamrakar et al., 2014a and Tamrakar et al., 2014c). Some additional peaks were present in the FTIR spectra of phosphors prepared by combustion synthesis. The absorption bands around 3400 cm^{-1} was due to O–H stretching and around 1500 cm^{-1} was due to C–O stretching (Guo et al., 2004a, Guo et al., 2004b, Guo et al., 2004c and Nakamoto, 2009). These bonds become weaker with increasing annealing temperature and nearly disappear after heating at 900 °C. This result indicates that there are no OH or CO groups in the Gd_2O_3 nanocrystals annealed at 900 °C (Fig. 2A–C). These bands were absent in FTIR spectra of Gd_2O_3 phosphor synthesized by solid state reaction method (Fig. 2D).

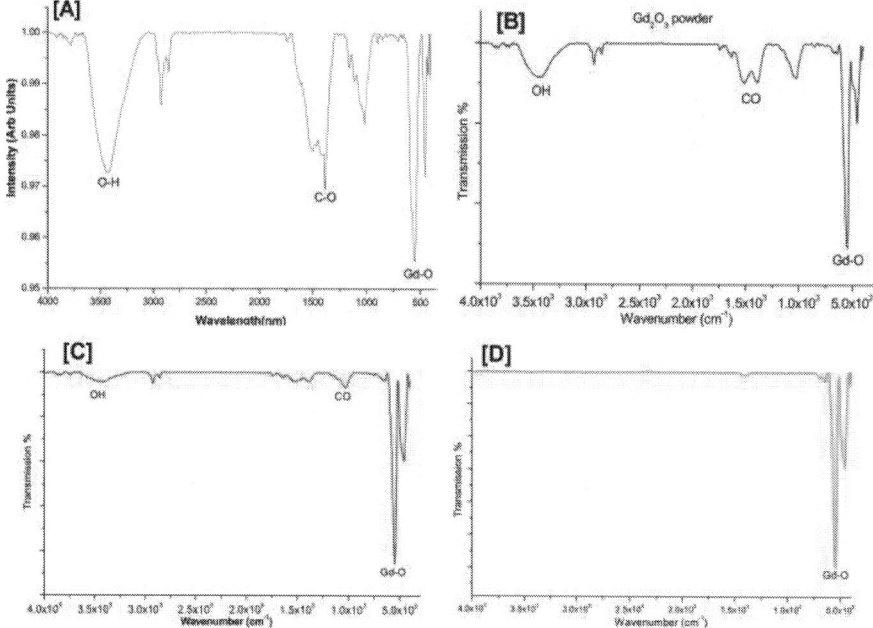

Figure 2. FTIR spectra of Gd_2O_3 (A) prepared by combustion method freshly prepared (B) annealed at 600 °C (C) annealed at 900 °C (D) prepared by solid state method.

Scanning Electron Microscope (SEM)

The morphology and topography of the samples were studied using imaging techniques such as Scanning Electron Microscopy (SEM). In SEM image of the phosphor prepared by solid state method showed the rough and spongy surface morphology is evident which makes it difficult to estimate the crystal size due to agglomeration of the particles. Whereas the phosphor prepared by combustion method is comparatively less compact (Fig. 3A–D).

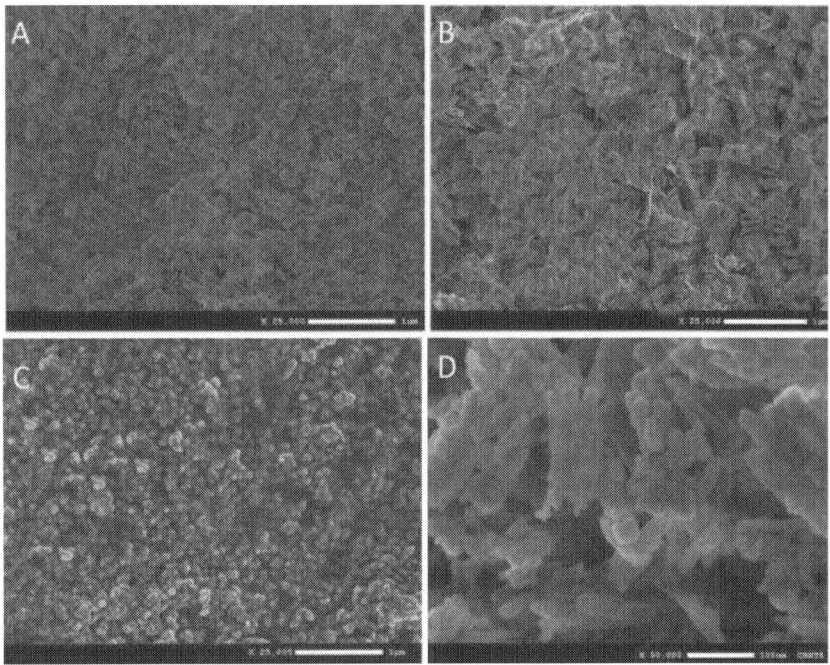

Figure 3. SEM of Gd_2O_3 (A) prepared by combustion synthesis method freshly prepared (B) annealed at 600 °C (C) annealed at 900 °C (D) prepared by solid state reaction.

Energy Dispersive X-Ray Analysis (EDX)

The EDX spectra represent elemental analysis of the samples prepared by both methods (Fig. 4). In the spectrum, intense peak of Gd and O are present which confirms the formation of pure Gd_2O_3 phosphor. The EDX of the Gd_2O_3 phosphor prepared by both methods were identical.

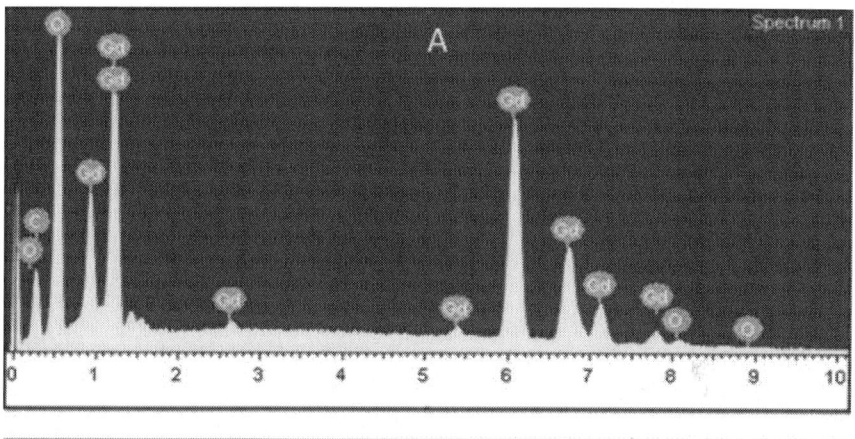

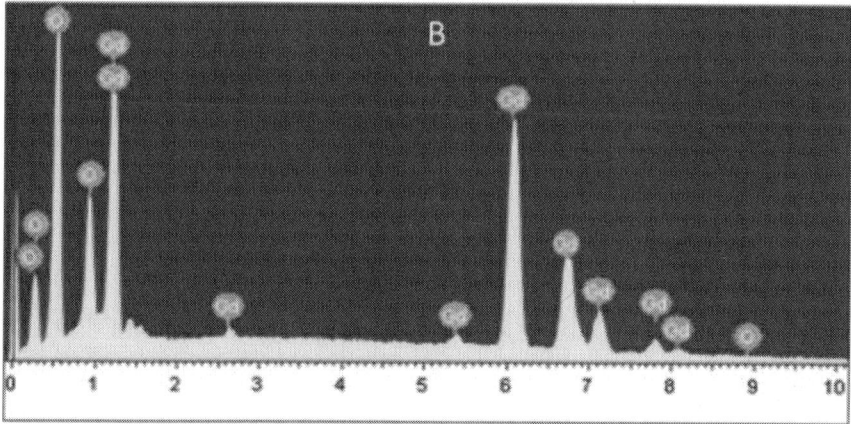

Figure 4. Energy dispersive X-ray analysis (A) Gd$_2$O$_3$ by solid state reaction (B) Gd$_2$O$_3$ by combustion synthesis method.

Transmission Electron Microscopy (TEM)

Morphology of Gd$_2$O$_3$ phosphor analyzed by the transmission electron microscopy (TEM) results. TEM analysis shows the prepared phosphors by both methods were in nano range. The exact particle sizes of phosphors by combustion method for freshly prepared, annealed at 600 °C and annealed at 900 °C according to TEM analysis were 7 nm, 12 nm and 22 nm (Fig. 5A). The TEM analysis show that the particle size for sample prepared by solid state method was 42–44 nm (Fig. 5B). The TEM results matches with the XRD and SEM results.

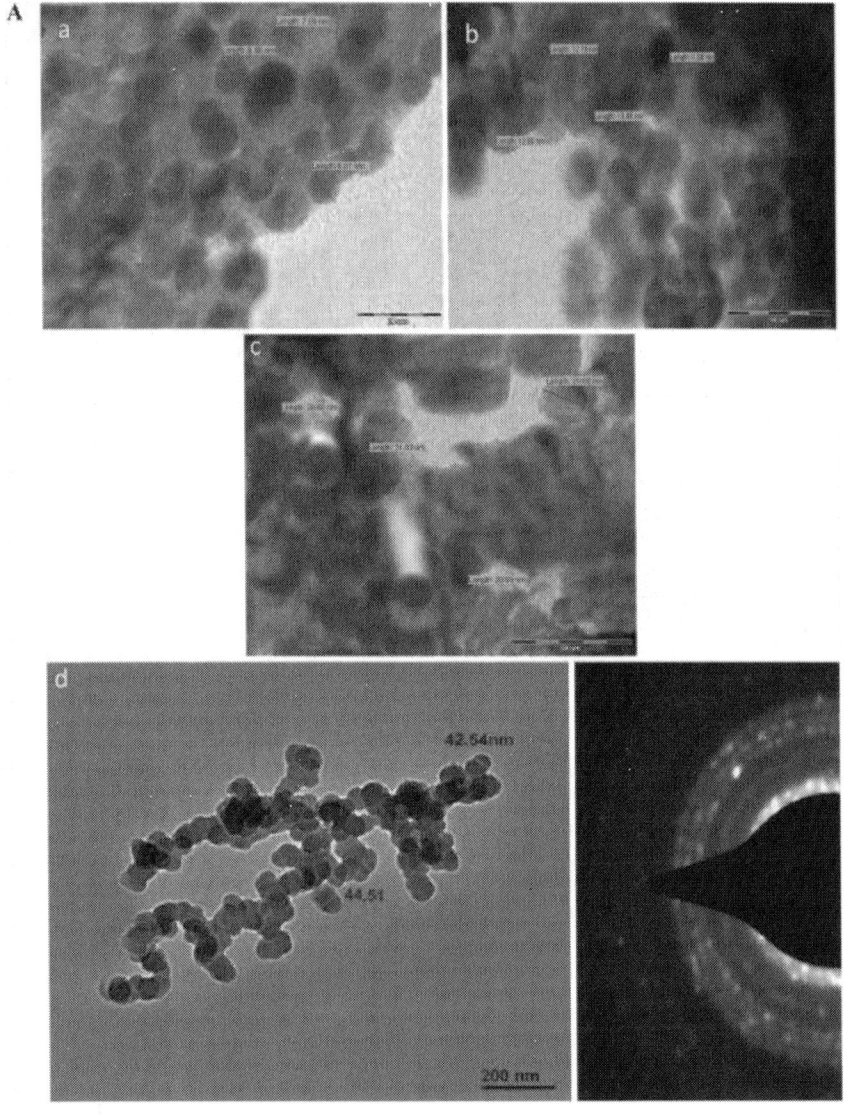

Figure 5. A) TEM image of the Gd_2O_3 (a) prepared by combustion synthesis method freshly prepared (b) annealed at 600 °C (c) annealed at 900 °C. (d) prepared by solid state reaction.

Photoluminescence Study

The photoluminescence excitation and emission spectra of the Gd_2O_3 phosphors prepared by both methods were recorded. The excitation spectrum was recorded under emission wavelength 515 nm. The excitation spectra have four peaks, positioned at, 230, 254, 260 nm and 275 nm (Fig. 6A and B). The 230 nm emission peak is attributed to the excitation band of the Gd_2O_3 host. In the excitation spectra, the peak at 275 nm is due to f–f transition of $^8S_{7/2} \rightarrow {}^6I_J$. The other weak excitation peaks at 230, 254 and 260 nm are attributed to the $^8S_{7/2} \rightarrow {}^6D_J$ (Bosze et al., 2003 and Singh et al., 2011). The room temperature emission spectra of Gd_2O_3 were recorded under 275 nm excitation. The emission spectra have peak at UV region in between 318 and 370 nm along with blue band around 468–494 nm, green around 515–568 nm and red emission at 616–625 nm. The emission lines at UV region at 312 was the intensive one and assigned due to $^6P_{7/2} \rightarrow {}^8S_{7/2}$ transition. The peaks in the visible region of the emission spectra may be due to transition from ^6G_J state which may be identified by the detailed energy level scheme for Gd^{3+} (Wegh, Donker, Meijerink, Lamminmaki, & Holsa, 1997). The other emission peaks in UV region were centered at 368 nm, 398 nm that may be due to radiative recombination of hole and electrons in Gd_2O_3 crystal. The emission peak in the blue region around 368 nm to 494 nm could be attributed to different types of surface defects of Schottky and Frenkel types. In oxide systems, typically oxygen vacancies and interstitials are vastly prevalent and the manifestation of such defects, contributes significantly to the modified photoluminescence response (Dhananjaya et al., 2012 and Lushchik et al., 2003). The green band in between 515 and 586 nm may be due to stark level transition from the ^6G_J state of Gd^{3+} ion. The red emission band around 616–625 nm corresponds to $^6G_J \rightarrow {}^6P_J$ transition (Fig. 7A and B).

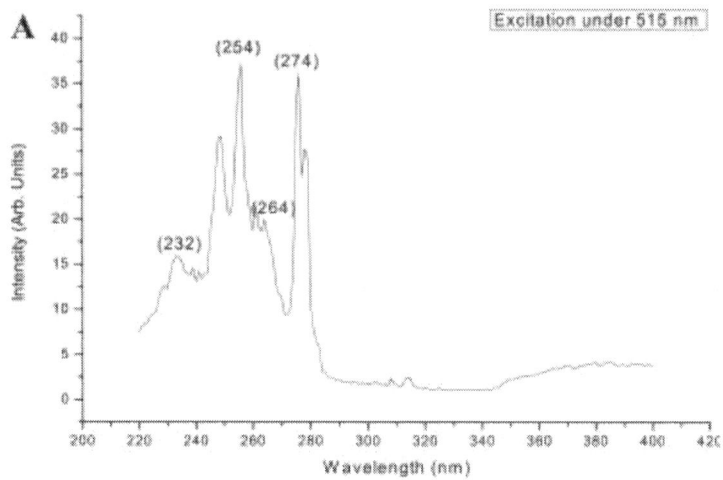

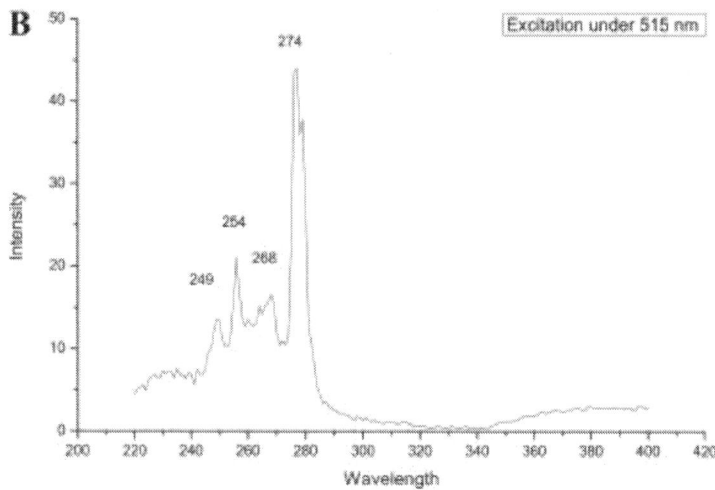

Figure 6. A) Excitation spectra of Gd_2O_3 phosphor by solid state reaction method. B) Excitation spectra of Gd_2O_3phosphor by combustion synthesis method.

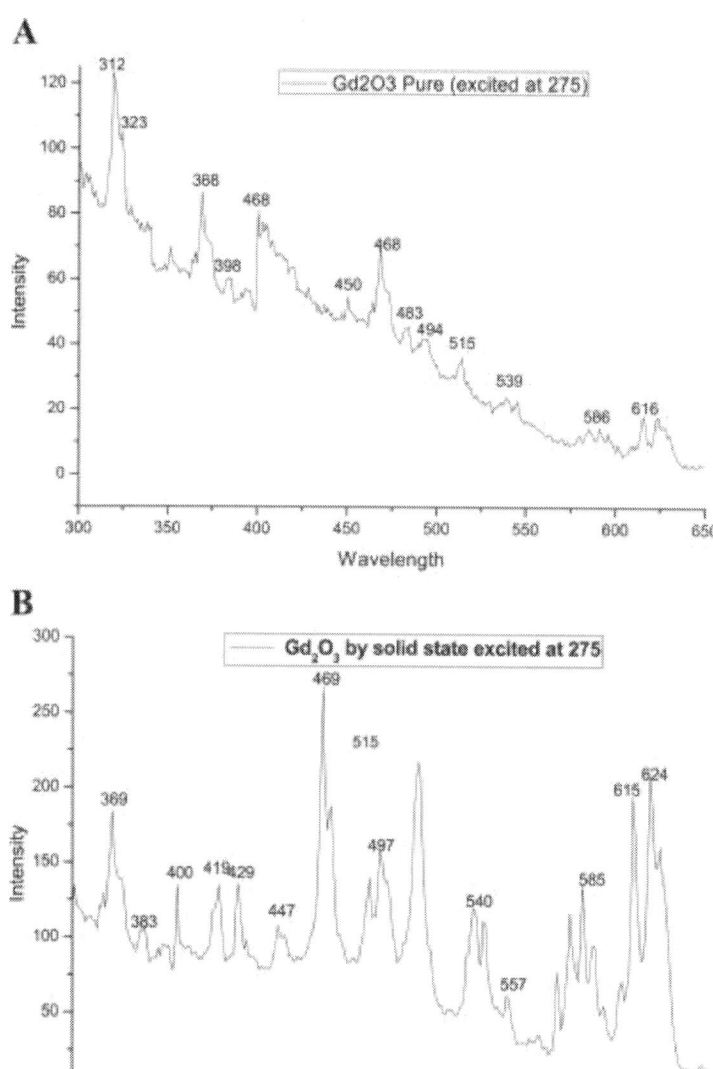

Figure 7. A) Emission spectra of Gd₂O₃ phosphor by combustion synthesis method. B) Emission spectra of Gd₂O₃phosphor by solid state reaction method.

Effect of Annealing on Emission Spectra of Gd₂O₃ Prepared by Combustion Synthesis

To observe the effect of heat on the photoluminescence behavior of the Gd₂O₃ phosphor the emission spectra were recorded after annealing the prepared phosphor under 600, 800, 900 °C for 2 h. The PL intensity becomes higher with increasing temperature of heat treatment for the phosphor prepared by combustion method (Fig. 8). The annealing temperature affects the crystalline behavior and phase of the phosphor. Crystalline behavior or particle size of the phosphor increases with increasing annealing temperature. The surface area to volume ratio decreases with increase in particle size of the phosphor, as a result the concentration of CO_2 and H_2O on surface decreases. The vibrations of CO_2 and H_2O destroy the energy gap, which results in decrease in luminescence behavior. Large concentration of CO_2 and H_2O on the surface of phosphor quenches the luminescence intensity of the phosphor (Capobianco et al., 2002 and Dhananjaya et al., 2012).The increasing annealing temperature decreases OH^- group's contamination in the prepared Gd_2O_3 phosphors above 800 °C. It is usual that the luminescence lifetime decreases significantly with decreasing hydroxyl contamination. However, luminescence quenching due to hydroxide group was not observed. Therefore, it can be concluded that there is no OH^- influence for quenching of luminescence in Gd_2O_3 annealed at 900 °C. For the phosphor prepared by solid state method, no effect of annealing was observed in the investigation of emission spectra.

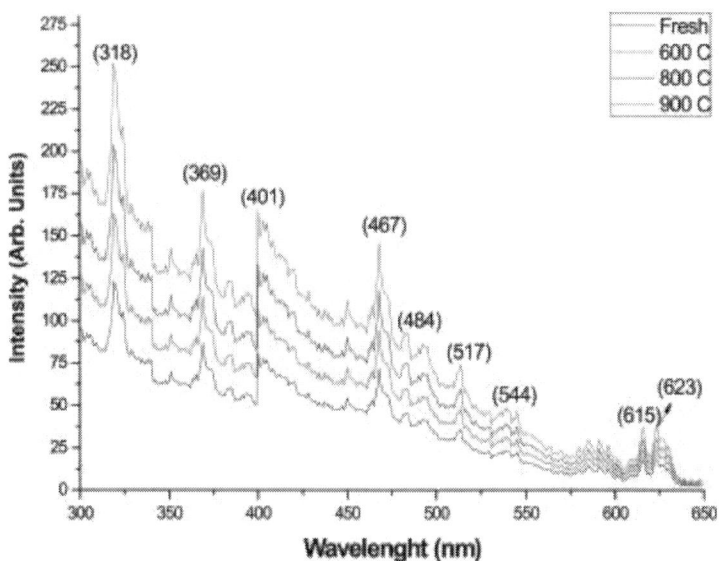

Figure. 8. Effect of annealing on emission spectra of Gd_2O_3 phosphor by combustion synthesis method.

Comparison of Emission Spectra of Solid State and Combustion Synthesis

Comparing the two synthesis, the overall shape of the emission spectra does not change (Fig. 9). However, the absolute emission intensity of the sample prepared by solid state reaction method is more than that of phosphor prepared by combustion synthesis. In fact the temperature involved during the synthesis affects the emission intensity of the phosphor. The sample prepared by the combustion synthesis required a further thermal treatment at 900 °C for 1 h after preparation to eliminate the residual unburnt species, therefore, the phosphor synthesized by solid state synthesis method were also treated at the same conditions for comparison. However, this step does not affect the intensity of the luminescence emission in the samples prepared by solid state method and it is not usually required in this method, since the high temperature developed during the synthesis is already sufficient to decompose the precursors and produce phosphors, eliminating the need for any further post synthesis heat treatment.

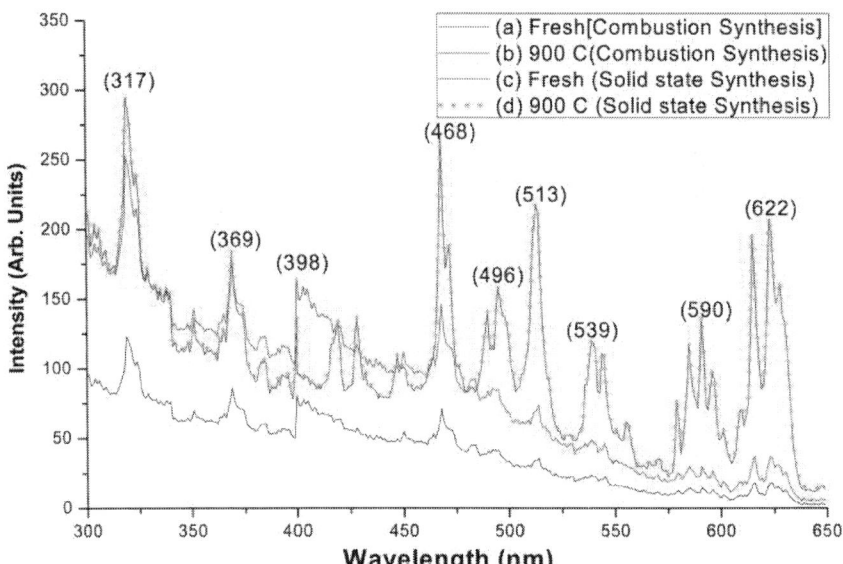

Figure 9. Comparison of Emission spectra of Gd_2O_3 phosphor by (a) combustion synthesis freshly prepared (b) combustion synthesis annealed at 900 °C (c) solid state reaction freshly prepared (d) solid state reaction annealed at 900 °C.

CONCLUSION

A comparison of the structural, morphological and upconversion properties of Gd_2O_3 nanocrystals prepared by solid state synthesis and combustion synthesis was carried out. The nanocrystal samples prepared by solid state showed greater uniformity and less aggregation, which are desired morphological properties in various applications, such as in the display field. The emission intensity following UV excitation was higher for the samples synthesized by solid state synthesis, without the requirement for further thermal treatment to eliminate the undecomposed residues. Low aggregation, uniformity and strong emission intensity, added to the possibility of varying the synthetic settings to modulate the properties of the final products, confirm the high potential of the solid state technique for the production of luminescence phosphor in one-step.

ACKNOWLEDGMENT

We are very grateful to NIT Raipur for XRD characterization and also thankful to Dr. Mukul Gupta for his co-operation. I am very thankful to SAIF, IIT, Bombay and IIT Delhi for other characterization such as SEM, TEM, FTIR and EDX. Very thankful to Dr. KVR Murthy M.S. University, Baroda (Chairman Luminescence Society of India) for PL study.

REFERENCES

1. Bhargave, R. N., & Gallagher, D. (1994). Optical properties of manganese doped nanocrystals of ZnS. Physical Review Letters, 72, 416.
2. Bosze, E. Z., Hirata, G. A., Shea-Rhower, L. E., & Mckittrick, J. (June 2003). Improving the efficiency of a blue-emitting phosphor by an energy transfer from Gd3þ to Ce3þ. Journal of luminescence, 104(1e2), 47e54.
3. Capobianco, J. A., Vetrone, F., Boyer, J. C., Speghini, A., & Bettinelli, M. (2002). Enhancement of red emission (4 F9/ 2 / 4 I15/2) via upconversion in bulk and nanocrystalline cubic Y2O3:Er3þ. Journal of Physical Chemistry B, 106, 1181.
4. Dhananjaya, N., Nagabhushana, H., Nagabhushana, B. M., Rudraswamy, B., Sharma, S. C., Sunitha, D. V., et al. (2012).

5. Effect of different fuels on structural, thermo and photoluminescent properties of Gd2O3 nanoparticles. Spectochimica Acta. Part A, 96(2012), 532e540.

6. Dhananjaya, N., Nagabhushana, H., Nagabhushana, B. M., Rudraswamy, B., Shivakumara, C., & Chakradhar, R. P. S. (2011). Effect of Liþ ion on enhancement of photoluminescence in Gd2O3:Eu3þ nanophosphors prepared by combustion technique. Journal of Alloys and Compounds, 509, 2368e2374.

7. Dosev, D., Kennedy, I. M., Godlewski, M., Gryczynski, I., Tomsia, K., & Goldys, E. M. (2006). Fluorescence upconversion in Sm-doped Gd2O3. Applied Physics Letters, 88, 11906e11909.

8. Garcia Murillo, A., Le Luyer, C., Dujardin, C., Martin, T., Garapon, C., Pedrini, C., et al. (2001). Elaboration and scintillation properties of Eu3þ-doped Gd2O3 and Lu2O3 solegel films. Nuclear Instruments and Methods in Physics Research Section A, 486(1e2), 181e185, 21 June 2002, 16, 39.

9. Grier, D., & Mccarthy, G. (1991). North Dakota state university. Fargo: North Dakota, USA, ICCD Grant-in-Aid. Guinier, A. (1963). X-ray diffraction. San Francisco, Calif, USA: Freeman.

10. Guo, H., Li, Y., Wang, D., Zhang, W., Yin, M., Lou, L., et al. (2004). Blue upconversion of cubic Gd2O3:Er produced by green laser. Journal of Alloys and Compounds, 376, 23e27.

11. Guo, H., Yang, X., Xiao, T., Zhang, W., Lou, L., & Mugnier, J. (2004). Structure and optical properties of solegel derived Gd2O3 waveguide films. Applied Surface Science, 230, 215e221.

12. Guo, H., Zhang, W., Yin, M., Lou, L., & Xia, S. (2004). Structure property and visible upconversion of Er3þ doped Gd2O3 nanocrystals. Journal of Rare Earths, 22(3), 365.

13. Korzenski, M. B., Lecoeur, Ph., Mercey, B., Camy, P., & Doualan, J. L. (2001). Preparation, patterning and luminescent properties of nanocrystalline Gd2O3:A (A¼Eu3þ, Dy3þ, Sm3þ, Er3þ) phosphor films via Pechini sol-gel soft lithography. Applied Physics Letters, 78, 1210.

14. Li, G. Z., Yu, M., Wang, Z. L., Lin, J., Wang, R. S., & Fang, J. (2006). Sol-gel fabrication and photoluminescence properties of SiO2 @ Gd2O3:Eu3þ coreeshell particles. Journal of Nanoscience and Nanotechnology, 6(5), 1416e1422.

15. Lushchik, A., Savikhin, F., & Tokbergenov, I. (May 2003). Electron and hole intraband luminescence in complex metal oxides. Journal of Luminescence, 102e103, 44e47.

16. Nakamoto, K. (2009). Infrared and Raman spectra of inorganic and coordination compounds. New Jersey: John Wiley & Sons Inc. ISBN: 10: 0471629790.

17. Pang, M. L., Lin, J., Fu, J.., Xing, R. B., Luo, C. X., & Han, Y. C. (2003). Preparation, patterning and luminescent properties of nanocrystalline Gd2O3:A (A¼Eu3þ, Dy3þ, Sm3þ, Er3þ) phosphor films via Pechini sol-gel soft lithography. Optical Materials, 23, 547e558.

18. Singh, V., Chakradhar, R. P. S., Rao, J. L., Leudox-Rak, I., & Kwak, H. (February 2011). Luminescence and EPR studies of Y2O3:Gd3þ phosphors prepared via solution combustion method. Journal of Materials Science, 46(4), 1038e1043.

19. Sun, L. D., Yao, J., Liu, C. H., Liao, C. S., & Yan, C. H. (2000). Rare earth activated nanosized oxide phosphors: synthesis and optical properties. Journal of Luminescence, 87e89, 447e450.

20. Tamrakar, R. K. (2012). Studies on absorption spectra of Mn doped CdS nanoparticles. LAP Lambert Academic Publishing, Verlag, ISBN 978-3-659-26222-7.

21. Tamrakar, R. K., & Bisen, D. P. (2013a). Optical and kinetic studies of CdS:Cu nanoparticles. Research on Chemical Intermediates, 39, 3043e3048.

22. Tamrakar, R. K., & Bisen, D. P. (2013b). Combustion synthesis and optical properties of ceria doped gadolinium oxide nano powder. AIP Conference Proceedings, 1536, 273. http://dx.doi.org/ 10.1063/1.4810206.

23. Tamrakar, R. K., Bisen, D. P., & Brahme, N. (2014). Characterization and luminescence properties of Gd2O3 phosphor. Research on Chemical Intermediates, 40, 1771e1779.

24. Tamrakar, R. K., Bisen, D. P., Robinson, C. S., Sahu, I. P., & Brahme, N. (2014). Ytterbium doped gadolinium oxide (Gd2O3:Yb3þ) phosphor: topology, morphology, and luminescence behaviour. Hindawi Publishing Corporation. Article ID 396147 Indian Journal of Materials Science, 7. Accepted 04.02.14.

25. Tamrakar, R. K., Bisen, D. P., Sahu, I. P., & Brahme, N. (2014). UV and gamma ray induced thermoluminescence properties of cubic Gd2O3:Er3þ phosphor. Journal of Radiation Research and Applied Sciences. http://dx.doi.org/10.1016/j.jrras.2014.07.003.

26. Tamrakar, R. K., Bisen, D. P., Upadhyay, K., & Tiwari, S. (2014). Synthesis and thermoluminescence behavior of ZrO2:Eu3þ with variable concentration of Eu3þ doped phosphor. Journal of Radiation Research and Applied Sciences. http://dx.doi.org/10. 1016/j.jrras.2014.08.006.

27. Tamrakar, R. K., Upadhyay, K., & Bisen, D. P. (2014). Gamma ray induced thermoluminescence studies of yttrium (III) oxide nanopowders doped with gadolinium. Journal of Radiation Research and Applied Sciences. http://dx.doi.org/10.1016/ j.jrras.2014.08.012.

28. Wang, F., Tan, W. B., Zhang, Y., Fan, X., & Wang, M. (2006). Luminescent nanomaterials for biological labelling. Nanotechnology, 17, R1eR13

29. Wegh, R. T., Donker, H., Meijerink, A., Lamminmaki, R. J., & Holsa, J. (1997). Vacuum-ultraviolet spectroscopy and quantum cutting for Gd3þ in LiYF4. Physical Review B, 56, 13841.

30. Yanhong, L., & Guangyan, H. (2007). Synthesis and luminescence properties of nanocrystalline Gd2O3:Eu3þ by combustion process. Journal of Luminescence, 124, 297e301.

31. Zou, Y., Tang, L., Cai, J. L., Lin, L. T., Cao, L.-W., & Meng, J.-X. (September 2014). Combustion synthesis and luminescence of monoclinic Gd2O3:Bi phosphors. Journal of Luminescence, 153, 210e214.

CITATION

Raunak Kumar Tamrakar, Durga Prasad Bisen, Nameeta Brahme, Comparison of photoluminescence properties of Gd2O3 phosphor synthesized by combustion and solid state reaction method, Journal of Radiation Research and Applied Sciences, Volume 7, Issue 4, October 2014, Pages 550-559, ISSN 1687-8507, http://dx.doi.org/10.1016/j.jrras.2014.09.005.

CHAPTER 4

Fabrication of Solid-state Thin-film Batteries using LiMnPO$_4$ thin films Deposited by Pulsed Laser Deposition

Daichi Fujimoto1, 2, Naoaki Kuwata1, Yasutaka Matsuda1, Junichi Kawamura1 and Feiyu Kang2

[1]Institute of Multidisciplinary Research for Advanced Materials, Tohoku University, Katahira 2-1-1, Aoba-ku, Sendai 980-8577, Japan

[2]Key Laboratory of Advanced Materials (MOE), School of Materials Science and Engineering, Tsinghua University, Beijing 100084, China

ABSTRACT

Solid-state thin-film batteries using LiMnPO$_4$ thin films as positive electrodes were fabricated and the electrochemical properties were characterized. The LiMnPO$_4$ thin films were deposited on Pt coated glass substrates by pulsed laser deposition. In-plane X-ray diffraction revealed that the LiMnPO$_4$ thin films were well crystallized and may have a texture with a (020) orientation. The deposition conditions were optimized; the substrate temperature was 600 °C and the argon pressure was 100 Pa. The electrochemical measurements indicate that the LiMnPO$_4$ films show charge and discharge peaks at 4.3 V and 4.1 V, respectively. The electrical conductivity of the LiMnPO$_4$ film was measured by impedance spectroscopy to be 2×10^{-11} S cm^{-1} at room temperature. The solid-state thin-film batteries that show excellent cycle stability were fabricated using the LiMnPO$_4$ thin film. Moreover, the chemical diffusion of the LiMnPO$_4$ thin film was studied by cyclic voltammetry. The chemical diffusion coefficient of the LiMnPO$_4$ thin film is estimated to be 3.0×10^{-17} cm^2 s^{-1}, which is approximately four orders magnitude smaller than the LiFePO$_4$ thin films, and the capacity of the thin-film battery was gradually increased for 500 cycles.

INTRODUCTION

The olivine-type $LiMPO_4$ (M = Fe, Mn, Co) proposed by Padhi et al. [1] has been demonstrated to be a promising cathode material for rechargeable Li batteries. Among the olivine-type materials, $LiMnPO_4$ has attracted attention because it has a higher working potential (4.1 V vs. Li/Li^+) than $LiFePO_4$ [2] and [3]. Unfortunately, the rate performance of the olivine-type $LiMnPO_4$ is limited by its poor electronic conductivity [4] and [5]. Although carbon coating [6] and [7] and particle size reduction [8], [9] and [10] have been demonstrated as effective strategies, further understanding of the intrinsic transport properties of $LiMnPO_4$ is required to reveal the factors that predominate the electrochemical properties. Thin-film electrodes are of great interest because they can serve as a simplified model to understand the electrochemical process of active materials [11]. Well-defined thin films are suitable for fundamental research since no binder and conductive additives are included. Moreover, thin-film cathode materials are also essential for all solid-state thin-film batteries (TFBs) [12], [13], [14] and [15]. Thin-film rechargeable Li batteries have numerous possible applications such as implantable medical devices, remote sensors, and smart cards [16], [17] and [18].

Numerous studies on the growth of olivine $LiFePO_4$ thin films by radio-frequency (RF) magnetron sputtering [19], [20] and [21], ion beam sputtering [22], pulsed laser deposition (PLD) [23], [24], [25], [26], [27], [28], [29] and [30], and electrostatic spray deposition (ESD) [31] have been reported. The $LiCoPO_4$ thin films have also been grown by RF magnetron sputtering [32], [33] and [34], ESD [35], and a sol–gel method [36]. However, very few studies on $LiMnPO_4$ thin films have been reported, possibly due to the difficulty of structure and composition control. Ma and Qin [31] have authored the report on $LiMnPO_4$ thin-films, in which they have reported the application of ESD combined with a sol–gel method. The films were deposited on a stainless steel substrate and then post-annealed at 600 °C to obtain crystalline $LiMnPO_4$. The charge and discharge characteristics were confirmed using liquid electrolytes. The reversible capacity was less than 8 mAh g^{-1}, which is 5% of the theoretical capacity; however, fabrication of solid-state TFBs using $LiMnPO_4$ films has not been reported. ESD films usually have rough surface morphologies, which may cause short-circuit problems in the battery. Thus, physical vapor deposition techniques are preferred to obtain a dense film with uniform thickness.

In this paper, we report the deposition of carbon-free $LiMnPO_4$ thin films on Pt coated SiO_2 glass substrates using a pulsed laser deposition (PLD) technique. One of the most important characteristics in PLD is the ability to realize stoichiometric transfer of ablated material from multicomponent targets. This arises from the non-equilibrium nature of the laser ablation process due to absorption of high laser energy density by a small volume of a target material. Since the $LiMnPO_4$ is a multicomponent oxide, PLD is suitable technique for thin-film deposition. Another advantage of PLD is the ability to operate with various background pressures of gases. Actually, the $LiMnPO_4$ thin films were sensitive to the process gases and the pressure, which were optimized to obtain the best electrochemical properties. The crystal structure, lattice vibration, surface morphology, and electrochemical performance of the $LiMnPO_4$ thin films were characterized. The solid–solid interface between the electrodes and the solid electrolyte is essential for the solid-state battery. A good contact between cathode and solid electrolyte is achieved by sequential PLD process using exchanging different target materials [37]. All solid-state TFBs were fabricated using the $LiMnPO_4$ film as the positive electrode. The TFBs were tested between 3.5 and 4.5 V vs. Li/Li^+, and demonstrated excellent cycle performance for 500 cycles.

EXPERIMENTAL DETAILS

The $LiMnPO_4$ powder was prepared by hydrothermal reaction at 150 °C for 12 h. The starting chemicals were lithium hydroxide monohydrate ($LiOH \cdot H_2O$), manganese acetate tetrahydrate ($Mn (CH_3COO)_2 \cdot 4H_2O$), and ammonium dihydrogen phosphate ($NH_4H_2PO_4$). The chemicals were placed into an autoclave with distilled water. The molar ratio of Mn and H_2O was 1:15, and the molar ratio of Li, Mn, and P in the precursor solution was set to 1.75:1.0:1.1. The $LiMnPO_4$ powder was ground by wet ball-milling with ethanol and then pressed into 1.66-mm-thick pellets with a diameter of 20 mm using a hydrostatic press. The pellet was then sintered at 800 °C for 10 h in Ar atmosphere. The relative density of the target was 97% of the theoretical density (3.44 g cm^{-3}) of $LiMnPO_4$. The composition ratio of the target was Li: Mn:P = 1.02:1.0:0.97 measured by inductive coupled plasma–optical emission spectrometry (ICP-OES, Perkin Elmer Optima 3300XL).

$LiMnPO_4$ thin films were grown on Pt/Cr/SiO$_2$ substrates (Sendai Sekiei Glass) by PLD. The fourth harmonic of a Nd: YAG laser (Spectra-Physics, LAB-150) was used with laser energy of 1.58 J cm^{-2}. The working pressure of Ar gas was fixed between 2 and 100 Pa after evacuating the chamber to 2×10^{-4} Pa, and the substrate temperature was set to 400, 500, 600, and 700 °C. The crystal structure of the $LiMnPO_4$ thin films was characterized by X-ray diffraction (XRD, Rigaku, Smart Lab 90TF) using Cu K$_\alpha$ radiation, and in-plane measurements with $2\theta\chi/\phi$ geometry were used. The lattice vibration of the film was analyzed by Fourier transform infrared (FTIR) spectroscopy (Perkin-Elmer, Spectrum GX). The FTIR spectra were measured by an attenuated total reflectance method. The surface morphology of the thin-films was observed by a field emission scanning electron microscope (FE-SEM, Hitachi S-4800). The operating voltage of the FE-SEM was 2.0 kV. The composition ratio of the $LiMnPO_4$ thin film was measured by ICP-OES analysis. The $LiMnPO_4$ film was dissolved with aqua regia. Reference solutions were made by diluting the Li, Mn and P standard solutions (Wako Pure Chemical).

The electrochemical properties of the $LiMnPO_4$ thin films were evaluated by cyclic voltammetry (CV) using a potentiostat (Bio-Logic, SP-150). A three-electrode cell was assembled in an Ar filled grove box using $LiMnPO_4$ thin films as the positive electrode. Lithium metal was used for the counter and reference electrodes, and 1 mol L^{-1} of LiPF$_6$was dissolved in ethylene carbonate–dimethyl carbonate (1:1 vol, Kishida Chemical) as the liquid electrolyte. The area of the $LiMnPO_4$ film was 0.56 cm^2, and the average thickness of the film was ca. 50 nm. The cell was tested by CV measurement in the range of 3.5–4.4 V vs. Li/Li$^+$ with a scan rate of 20 mV min^{-1}.

All solid-state TFBs, which consisted of Li/Li$_3$PO$_4$/LiMnPO$_4$, were deposited on a Pt/Cr/SiO$_2$ substrate. The conditions under which the LiMnPO$_4$ layer was deposited onto the Pt/Cr/SiO$_2$ substrate were 600 °C substrate temperature in Ar atmosphere of 100 Pa for 1 h using Nd: YAG laser. The amorphous Li$_3$PO$_4$ solid electrolyte was deposited at room temperature in O$_2$ atmosphere at 0.2 Pa for 3 h using an ArF excimer laser, and the Li film was deposited by thermal evaporation. Details of the Li$_3$PO$_4$ and Li depositions have been reported in a previous paper [15]. The area and the thickness of LiMnPO$_4$ film were 0.09 cm^2 and 50 nm, respectively. The TFBs were placed in a vacuum-tight stainless steel cell. Gold wire and silver paste were used to connect the cell and batteries electrically. CV measurement was performed at 25 °C under vacuum in the range of 3.5–4.5 V vs. Li/Li$^+$ with a scan rate of 20 mV min^{-1}, 50 mV min^{-1}, 1 mV s^{-1}, 2 mV s^{-1}, and 5 mV s^{-1} for 50 cycles. The cycle performance of the TFB was investigated for 500 cycles with a scan rate of 20 mV min^{-1}.

RESULTS AND DISCUSSION

Fig. 1 shows the in-plane XRD patterns of the LiMnPO$_4$ thin films deposited at various substrate temperatures. The Ar pressure was maintained at 100 Pa, and the film deposited at 400 °C showed three peaks due to Pt reflections from the Pt/Cr/SiO$_2$substrate. There is no Bragg reflection of LiMnPO$_4$, i.e., the LiMnPO$_4$ deposited at 400 °C was in the amorphous state. The LiMnPO$_4$ thin films deposited at 500 and 600 °C showed that several peaks that can be attributed to the Bragg reflections of crystalline LiMnPO$_4$. However, the film deposited at 700 °C indicated the decomposed phase which was possibly attributed to the manganese oxide compounds such as MnO$_2$ or Mn$_2$O$_3$. Thus, the crystalline, single phase LiMnPO$_4$ thin films are grown at substrate temperatures between 500 and 600 °C. The composition ratio of the LiMnPO$_4$ thin film prepared at 600 °C and 100 Pa of Ar was Li: Mn: P = 1.0 ± 0.2: 1.0: 0.88 ± 0.08 measured by ICP-OES analysis. The composition of thin film was close to the stoichiometry. The manganese was slightly higher than the phosphorous in the thin film. This result suggests the existence of manganese oxide in the thin film.

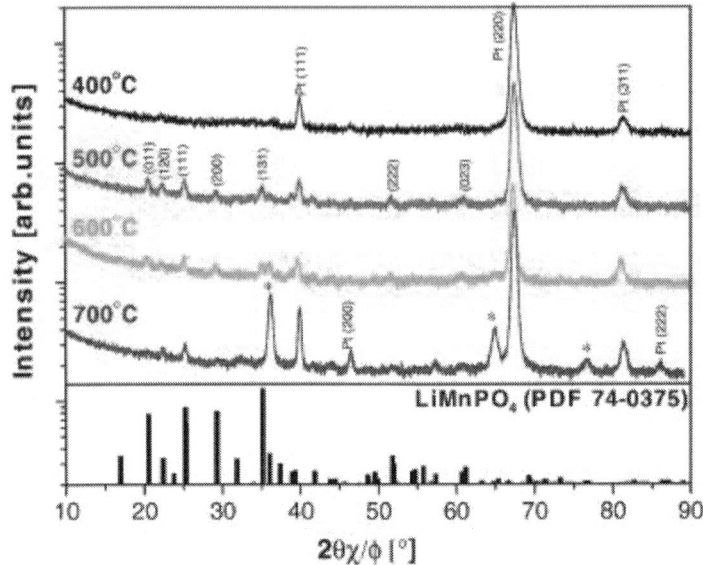

Figure 1. In-plane X-ray diffraction patterns for LiMnPO₄ thin films deposited at substrate temperatures of 400, 500, 600, and 700 °C at 100 Pa Ar pressure. Asterisk (*) shows peaks probably due to manganese oxide.

Fig. 2 shows the FTIR spectra of the LiMnPO₄ thin films deposited at various substrate temperatures. The films deposited at room temperature and 400 °C show broad peaks in the frequency region of 1000–1200 cm^{-1}. These peaks are attributed to the asymmetric stretching vibration (v_3) mode of the PO₄ group [38]. The broadening of the peaks indicates that the films were amorphous. The films deposited at 500 and 600 °C show sharp bands in the frequency of 1000–1200 cm^{-1}, which corresponds to the v_3 mode of the PO₄ group in the LiMnPO₄ crystal [38]. The thin film deposited at 700 °C showed the additional phase at 720 cm^{-1} that may be attributed to the stretching vibration of the MnO₆ octahedral in MnO₂ [39].

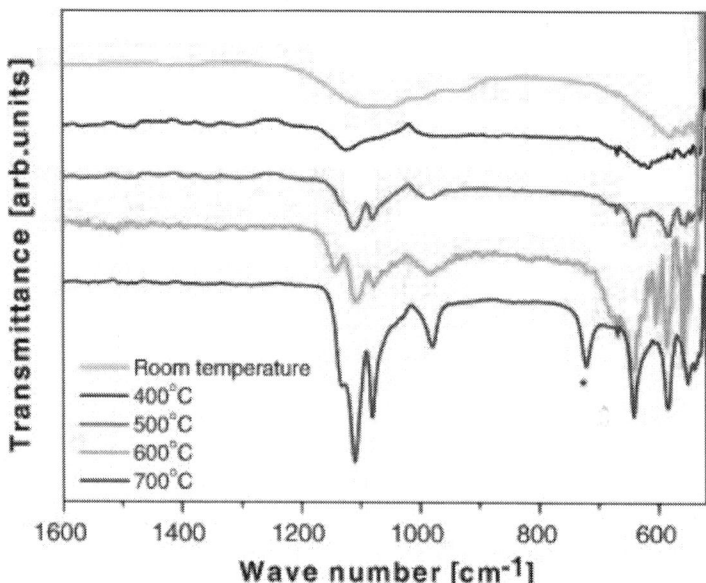

Figure 2. FTIR spectra for LiMnPO$_4$ thin films at different substrate temperatures at 100 Pa Ar pressure. Asterisk (*) shows a peak from manganese oxide.

Fig. 3 shows the in-plane XRD pattern of the LiMnPO$_4$ thin films deposited at various Ar pressures at 600 °C. All the films show Bragg peaks corresponding to the reflections of the LiMnPO$_4$. Small manganese oxide peaks were found in the films grown at lower Ar pressure. In addition, the intensity of the (020) peak located at 16.9° was different for the film deposited in 100 Pa, while the films deposited under the pressure of 2–50 Pa show the (020) diffraction peak. The low intensity of the (020) peak suggests that the film grown at 100 Pa may be textured. The small intensity of the (020) plane reflection along the in-plane suggests that the (020) plane can be parallel to the substrate. Because the Li ions in LiMnPO$_4$ are aligned along the b-axis, the orientation with the (020) plane parallel to the substrate can provide easier lithium diffusion along the b-axis.

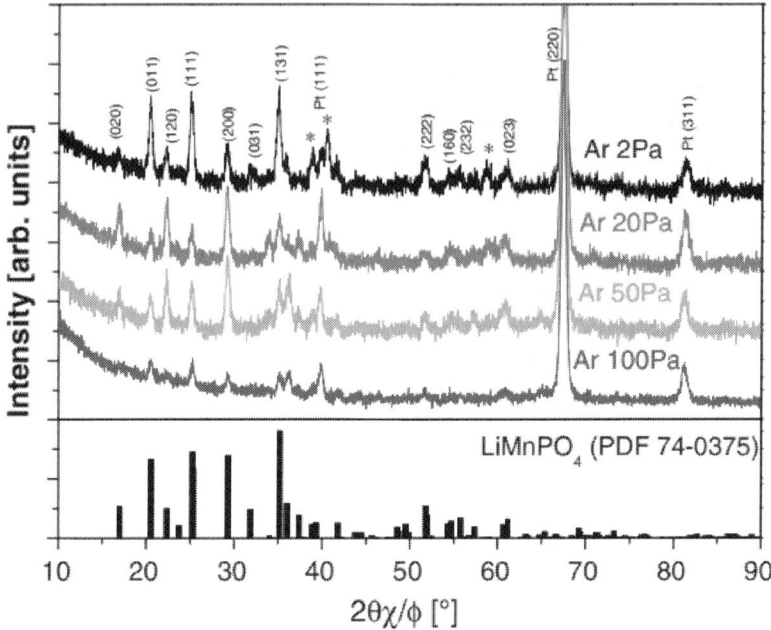

Figure 3. In-plane X-ray diffraction patterns for LiMnPO$_4$ thin films under different atmospheric pressures at a substrate temperature of 600 °C. Asterisk (*) shows peaks probably due to manganese oxide.

The surface morphology of the LiMnPO$_4$ films deposited in various Ar gas pressures is shown in Fig. 4. The film shows a smooth surface with small crystalline grains, and the size of crystalline grains was less than 100 nm. The grain size of the film deposited at 2 Pa Ar was small. It was less than 50 nm. The surface of the film deposited at 100 Pa Ar shows uniform grains with the grain size around 100 nm.

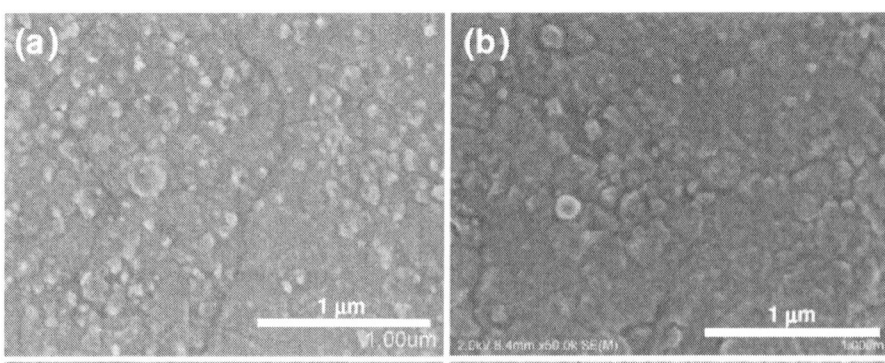

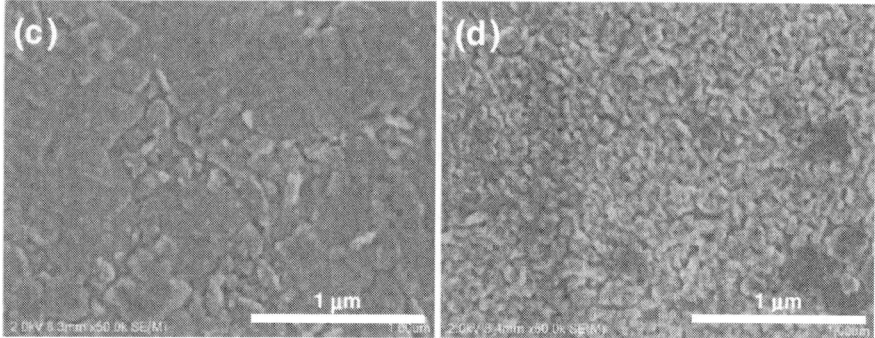

Figure 4. FE-SEM images of LiMnPO₄ thin films at different atmospheric pressures at a substrate temperature of 600 °C. (a) 2 Pa, (b) 20 Pa, (c) 50 Pa, and (d) 100 Pa.

Electrochemical properties of the LiMnPO₄ thin films fabricated under various conditions were evaluated by CV using a liquid electrolyte in the three-electrode cell. Fig. 5 shows the CV curves of the films grown at different substrate temperatures in 100 Pa Ar pressure. No intercalation (discharge) peak was observed in the film deposited at 400 °C due to the amorphous nature of the film. The film deposited at 500 °C showed a broad intercalation peak at ca. 4.0 V. The maximum intercalation peak at 4.0 V due to Mn^{2+}/Mn^{3+} redox was observed in the film deposited at 600 °C. The film deposited at 700 °C shows a small discharge peak probably due to the decomposition of LiMnPO₄ as observed in the XRD. The electrochemical properties of LiMnPO₄ films were found to be strongly dependent on the substrate temperature. At low temperature crystallization is not sufficient, and at high temperature manganese oxides are formed by decomposition. Such behavior has also been observed in LiFePO₄ films that show optimum capacity between 500 and 600 °C [21] and [23].

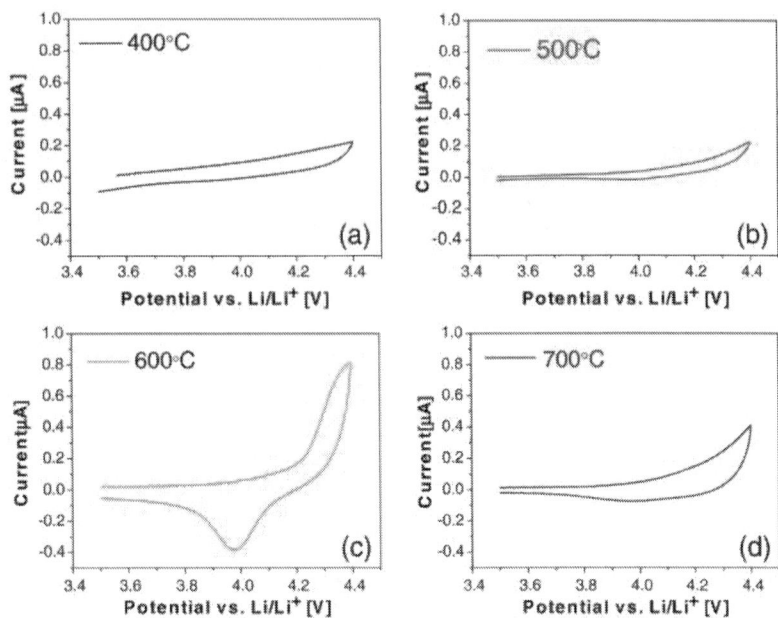

Figure 5. Cyclic voltammograms of the films deposited at (a) 400 °C, (b) 500 °C, (c) 600 °C, and (d) 700 °C at 100 Pa Ar pressure. The CV scan rate was 20 mV min^{-1}. The films were deposited on Pt coated SiO$_2$ glass substrates.

To calculate the intercalation capacity (negative current) and deintercalation capacity (positive current) of the films, the CV curve is integrated and divided by the scan rate $v(Vs^{-1})$ as follows:

$$Q = \frac{1}{v} \int_{V_1}^{V_2} i \, dV, \qquad (1)$$

where Q is the charge (C), i is the current (A), and V is the potential (V). The capacity (Ah) was calculated from the charge Q using a relationship 1 Ah = 3600 C. In the case of thin-film electrodes, the capacities obtained from CV measurements agree well to the capacities obtained from constant current measurements. The weight of the thin film was measured by an ultra-microbalance (XP2U, Mettler Toledo) to calculate the specific capacity (mAh g^{-1}). The typical weight of LMP thin film was about 10 µg when the film has the thickness of about 50 nm with the area of 0.56 cm^2. The experimental errors for the weight measurements were about ± 1 µg. Thus the errors for the calculated specific capacity were about ± 10%.

The charge and discharge capacities of the $LiMnPO_4$ film deposited at 600 °C were 34 ± 4 and 22 ± 2 mAh g^{-1}, respectively. The reversible capacity 22 ± 2 mAh g^{-1} is $13 \pm 2\%$ of the theoretical capacity (170 mAh g^{-1}) of $LiMnPO_4$. Although the reversible capacity was smaller than the capacity of the carbon-coated $LiMnPO_4$ powder sample (150 mAh g^{-1}) [3], it was comparable or slightly higher than that of previously reported $LiMnPO_4$ films deposited by ESD [31]. Generally, the electrochemical performance of pure $LiMnPO_4$ is explained by the intrinsically poor electronic and ionic conductivities of $LiMnPO_4$ [4] and [5].

The electronic conductivity of the $LiMnPO_4$ thin film was measured by impedance spectroscopy using comb-type Pt electrodes. The real and imaginary parts of the impedance spectra were shown in Fig. 6(a) and (b). The film resistance can be obtained by selecting the value of the real part (Z') at the frequency where the imaginary part (− Z'') goes through a local minimum and the real part (Z') shows a plateau region, which is indicated by black arrows in Fig. 6(a) and (b). Two plateaus were observed at high frequency (region 1) and the low frequency (region 2). The plateau at region 2 was attributed to the electronic conductivity, because the Pt electrodes act as the ion-blocking and electron-reversible electrodes at low frequency. The plateau at high frequency region 1 may be explained as the electronic conductivity in the grain or the ionic conductivity. The blocking behavior can be explained by the grain boundary or the lithium ion-blocking Pt electrodes. The activation energy of conductivity was estimated from Arrhenius plot of conductivity as shown in Fig. 6(c). The activation energy of the conductivity region 1 (0.59 eV) agrees well with that of the region 2 (0.60 eV). The similar activation energy suggests that both conductivities come from the electronic conductivity. Another possibility is the cooperative motion of electrons and lithium ions. The electronic conductivity estimated from the region 2 was 2×10^{-11} S cm^{-1} at room temperature. Assuming that the thickness of thin film is 100 nm, the resistance of the film for 1 cm^2 area becomes 5×10^5 Ω. If the current density is 10^{-6} A, the IR drop is estimated to be 0.5 V. Therefore, the electronic resistance does not limit the electrochemical reactions.

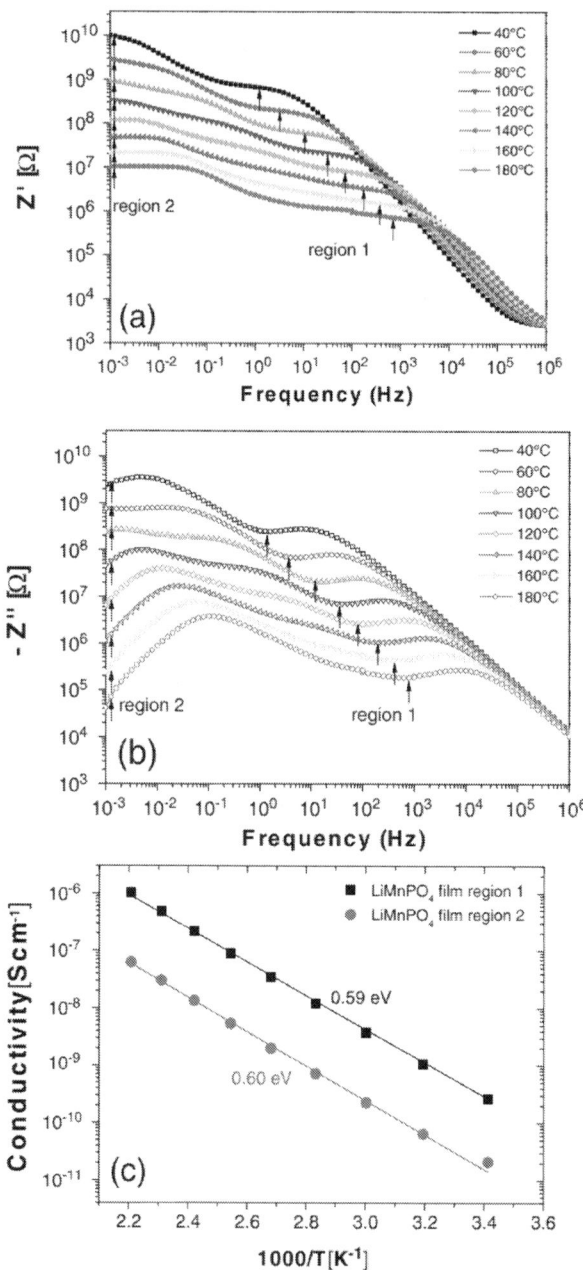

Figure 6. The conductivity of the LiMnPO$_4$ thin film prepared at 600 °C at 100 Pa Ar pressure. The temperature dependence of (a) real and (b) imaginary part of the impedance spectra is shown. The Arrhenius plot of the conductivity is shown in (c). The two conductivities are determined from regions 1 and 2 of the impedance spectra.

The electronic conductivity of the bulk $LiMnPO_4$ showed different values in the literature. Wang et al. reported the electronic conductivity of 1×10^{-9} Scm^{-1} at 298 K [40]; Delacourt et al. reported much lower value which was 3×10^{-9} Scm^{-1} at 573 K [4]. The electronic conductivity of the $LiMnPO_4$ thin film was 2×10^{-11} Scm^{-1} at 298 K measured in this study. Probably, the electronic conductivity is affected by the different preparation procedure. The electronic conductivity of the $LiMnPO_4$ depends on the electronic species (holes in Mn^{3+} or electrons in Mn^+) accompanied with lithium or oxygen vacancies [41]. Thus, the slight deviation from the stoichiometry may cause the large difference in the electronic conductivity.

Fig. 7 shows the CV curves of the film grown at 600 °C with different Ar pressures. The discharge peak due to Mn^{2+}/Mn^{3+} redox in $LiMnPO_4$ was observed in all films. The discharge peaks gradually increased with increasing Ar pressure and attained a maximum at 100 Pa. The high capacity of the film grown at 100 Pa may be explained by different crystallographic properties. The in-plane XRD measurements reveal that the film prepared at 100 Pa showed a preferred texture, as illustrated in Fig. 3, where the (020) plane can be directed to the out-of-plane to the substrate. Because the ionic motion in $LiMnPO_4$ is strongly anisotropic, this pronounced texture may increase lithium diffusion along the b-axis in the films grown in 100 Pa.

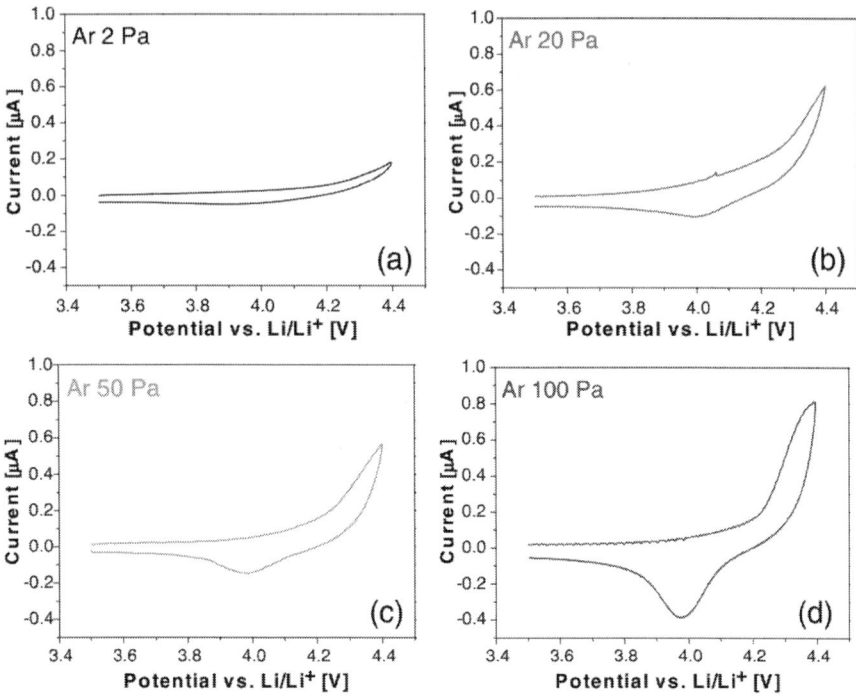

Figure 7. Cyclic voltammograms of the films deposited in Ar at (a) 2 Pa, (b) 20 Pa, (c) 50 Pa, and (d) 100 Pa. The CV scan rate was 20 mV min⁻¹. The films were deposited on Pt coated SiO_2 glass substrates.

All solid-state TFBs were fabricated by depositing $LiMnPO_4$ as the positive electrode, amorphous Li_3PO_4 as the solid electrolyte, and Li film as the negative electrode on a $Pt/Cr/SiO_2$ substrate. The $LiMnPO_4$ film was deposited under optimized conditions: 600 °C substrate temperature and 100 Pa Ar pressure. The thickness of the $LiMnPO_4$ film was ca. 50 nm, and the active area was 0.09 cm^2. The weight of the active material was expected to be 1.6 µg. Fig. 8(a) shows the CV curves of the TFB, and the extraction/insertion peaks due to the Mn^{2+}/Mn^{3+} redox of $LiMnPO_4$ was observed at ca. 4.3 V and 4.0 V, respectively. Excellent reversibility was achieved by good contact between $LiMnPO_4$ and the amorphous Li_3PO_4 electrolyte as well as the $LiCoO_2$ and $LiCoMnO_4$ films [15] and [42]. In addition, no decomposition of the electrolyte was observed from 3.5 to 4.5 V due to the large electrochemical window of the solid electrolyte. The stable electrochemical performance is suitable for the fundamental study of the model system of $LiMnPO_4$.

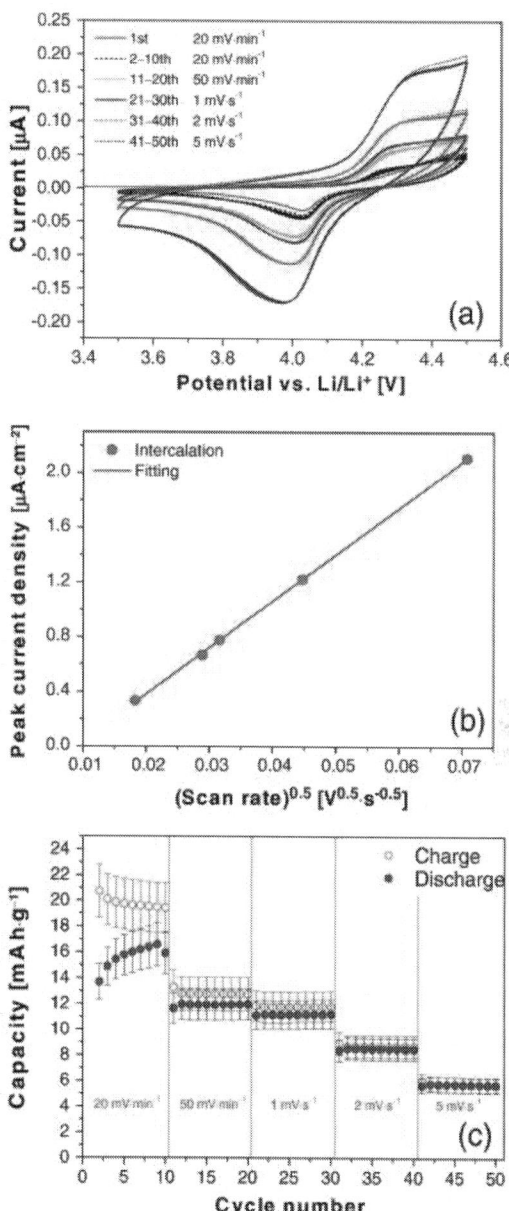

Figure 8. CV of all solid-state thin-film batteries Li/Li$_3$PO$_4$/LiMnPO$_4$; (a) CV curves at various scan rates and (b) Randles–Sevcik plot obtained from intercalation peak of the LiMnPO$_4$ film.

Usually, the LiPON electrolyte is known as a solid electrolyte with better stability than Li_3PO_4. However, the Li_3PO_4 thin film prepared by PLD using ArF excimer laser showed good stability and ionic conductivity [15]. The electrochemical stability of Li_3PO_4 film prepared by PLD is greater than 4.7 V, while the Li_3PO_4 film prepared by RF sputtering is decomposed at 3.6 V [43]. Pyrophosphate groups with POP bonds included in the Li_3PO_4 film prepared by PLD may improve the electrochemical stability.

The chemical diffusion of the $LiMnPO_4$ film was studied by CV measurements. The CV curves were obtained at different scan rates, as shown in Fig. 8(a). According to the Randles–Sevcik equation, the maximum current density i_p (A cm^{-2}) of the intercalation/deintercalation peak depends on the square root of the chemical diffusion coefficient $\tilde{D}$ (cm^2 s^{-1}) [44] and [45]

$$i_p = 0.4463 \cdot zFC \cdot \sqrt{\frac{zFv\tilde{D}}{RT}}.$$

(2)

Here, z denotes the valence of the mobile ions, F is Faraday's constant (C mol^{-1}), C denotes the concentration of mobile ions (mol cm^{-3}), v denotes the scan rate (V s^{-1}), R denotes the gas constant (J K^{-1} mol^{-1}), and T is temperature (K). According to Eq. (2), plotting i_p as a function of $\sqrt{v}$ should result in a straight line with a slope proportional to $\sqrt{\tilde{D}}$. For the $LiMnPO_4$ film, a Randles–Sevcik plot is shown in Fig. 8(b).

As expected, i_p depends linearly on $\sqrt{v}$, indicating a diffusion controlled process. Therefore, the chemical diffusion coefficient is calculated for the intercalation process and is found to be $\tilde{D} = 3.0 \times 10^{-17}$ cm^2 s^{-1}. This chemical diffusion coefficient value of $LiMnPO_4$ shows good agreement with the value of carbon-coated $LiMnPO_4$nanocrystals, which was between 10^{-16} and 10^{-17} obtained by electrochemical impedance spectroscopy [3]. In contrast, the $LiFePO_4$ thin film having the same olivine structure shows chemical diffusion coefficient of approximately

10^{-13} cm^2 s^{-1}; this value was obtained by CV [22]. The chemical diffusion coefficient of the LiMnPO$_4$ thin films is four orders of magnitude smaller than the value of the LiFePO$_4$ thin film. The relatively low charge and discharge current and capacity found in this study can be explained by the low chemical diffusion coefficient of the LiMnPO$_4$ thin film.

Finally, the cycle stability of the thin-film battery is shown for 500 cycles. Fig. 9 (a) shows the CV curves and Fig. 9(b) shows the specific capacity for 500 cycles. The CV curves inFig. 8(a) clearly show the excellent cycle stability of the thin-film battery using LiMnPO$_4$film as the positive electrode. This cycle stability results from the rigid structure of olivine-type LiMnPO$_4$ with a PO$_4$ framework. The specific capacity after 500 cycles was 28 ± 3 mAh g^{-1} and the volumetric capacity was 10 ± 1 μAh cm^{-2} μm^{-1}, which is $16 \pm 2\%$ of the theoretical capacity. After long term cycling, the capacity increased gradually. The unusual behavior of the capacity is due to the gradual increase of the utilization of LiMnPO$_4$. One possible explanation for this behavior is the crystallization of the remaining amorphous part, which increases the utilization of the LiMnPO$_4$ film. The other explanation is the formation of Li vacancies and electronic species (Mn^{3+}), during charge–discharge cycles. The Li vacancies and electronic species enhance the electronic conductivity of the LiMnPO$_4$. Since the utilization of LiMnPO$_4$ was limited by slow chemical diffusion coefficient, the increase in the electronic conductivity improves the utilization of LiMnPO$_4$.

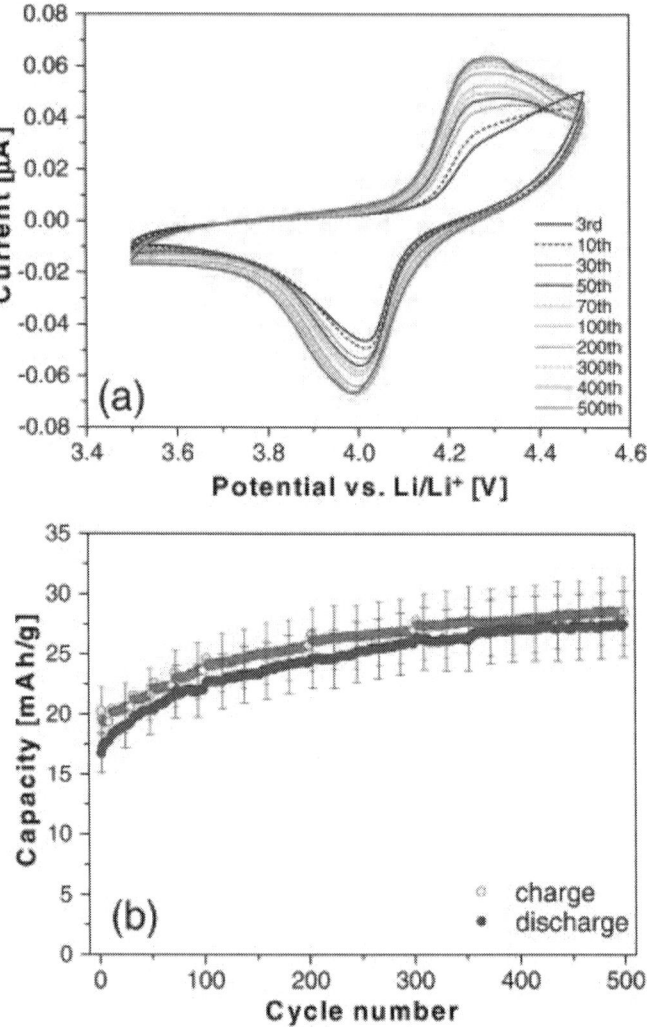

Figure 9. CV of thin-film battery Li/Li$_3$PO$_4$/LiMnPO$_4$ for 500 cycles at 20 mV min^{-1}; (a) CV curves and (b) specific capacities of the LiMnPO$_4$ film.

CONCLUSION

An olivine-type LiMnPO$_4$ thin-film cathode was fabricated by PLD using the stoichiometric LiMnPO$_4$ target. Substrate temperature and Ar gas pressure during deposition strongly affect the film structure and electrochemical properties. A weak (020) texture observed in the in-plane XRD provides a favorable orientation for lithium diffusion in the film. The

$LiMnPO_4$ thin film prepared under optimized conditions shows charge and discharge peaks at ca. 4.1 V due to Mn^{2+}/Mn^{3+} redox of $LiMnPO_4$, and the reversible capacity was 22 ± 2 mAh g^{-1} using the liquid electrolyte. The electronic conductivity of the $LiMnPO_4$ film was 2×10^{-11} S cm^{-1} at room temperature. All solid-state TFBs were fabricated using $LiMnPO_4$ film and an amorphous Li_3PO_4 electrolyte. The TFBs show the redox peak at ca. 4.1 V with excellent cycle stability. CV was applied to study the chemical diffusion of the $LiMnPO_4$ thin film using the solid-state battery. The chemical diffusion coefficient of the $LiMnPO_4$ thin film is estimated to be 3.0×10^{-17} cm^2 s^{-1}, which is about four orders magnitude smaller than the $LiFePO_4$ thin films. After long term cycling, i.e., 500 cycles, the capacity gradually increased to 28 ± 3 mAh g^{-1}. This behavior suggests the gradual increase in the utilization of the $LiMnPO_4$ thin film. The work on thin-film batteries using the other olivine materials $LiMPO_4$ (M = Fe, Co, Ni) are under investigation.

ACKNOWLEDGMENTS

This research was conducted under a Joint Education Program between Tsinghua University and Tohoku University. This work was partly supported by the Japan Science and Technology Agency, Advanced Low Carbon Technology Research and Development Program, "Next-generation Rechargeable Battery Program." The in-plane XRD measurement was supported by "Tohoku Innovative Materials Technology Initiatives for Reconstruction" funded by the Ministry of Education, Culture, Sports, Science and Technology and the Reconstruction Agency, Japan.

REFERENCES

1. A.K. Padhi, K.S. Nanjundaswamy, J.B. Goodenough, Phospho-olivines as positiveelectrode materials for rechargeable lithium batteries, J. Electrochem. Soc. 144 (1997) 1188.
2. D. Choi, D.Wang, I.-T. Bae, J. Xiao, Z. Nie, W. Wang, V.V. Viswanathan, Y. Jung Lee, J.-G. Zhang, G.L. Graff, Z. Yang, J. Liu, LiMnPO4 nanoplate

grown via solid-state reaction in molten hydrocarbon for Li-ion battery cathode, Nano Lett. 10 (2010) 2799.

3. H.-C. Dinh, S.-I. Mho, Y. Kang, I.-H. Yeo, Large discharge capacities at high current rates for carbon-coated LiMnPO4 nanocrystalline cathodes, J. Power Sources 244 (2013) 189.

4. C. Delacourt, L. Laffont, R. Bouchet, C. Wurm, J.-B. Leriche, M. Morcrette, J.-M. Tarascon, C. Masquelier, toward understanding of electrical limitations (electronic, ionic) in LiMPO4 (M = Fe, Mn) electrode materials, J. Electrochem. Soc. 152 (2005) A913.

5. K. Rissouli, K. Benkhouja, J.R. Ramos-Barrado, C. Julien, Electrical conductivity in lithium orthophosphates, Mater. Sci. Eng. B 98 (2003) 185.

6. S. Moon, P. Muralidharan, D.K. Kim, Carbon coating by high-energy milling and electrochemical properties of LiMnPO4 obtained in polyol process, Ceram. Int. 38 (2012) S471.

7. S.K. Martha, B. Markovsky, J. Grinblat, Y. Gofer, O. Haik, E. Zinigrad, D. Aurbach, T. Drezen, D. Wang, G. Deghenghi, I. Exnar, LiMnPO4 as an advanced cathode material for rechargeable lithium batteries, J. Electrochem. Soc. 156 (2009) A541.

8. T.-H. Kim, H.-S. Park, M.-H. Lee, S.-Y. Lee, H.-K. Song, Restricted growth of LiMnPO4 nanoparticles evolved from a precursor seed, J. Power Sources 210 (2012) 1.

9. T. Drezen, N.-H. Kwon, P. Bowen, I. Teerlinck, M. Isono, I. Exnar, Effect of particle size on LiMnPO4 cathodes, J. Power Sources 174 (2007) 949.

10. K. Dokko, T. Hachida, M. Watanabe, LiMnPO4 nanoparticles prepared through the reaction between Li3PO4 and molten aqua-complex of MnSO4, J. Electrochem. Soc. 158 (2011) A1275.

11. J. Wang, X. Sun, Understanding and recent development of carbon coating on LiFePO4 cathode materials for lithium-ion batteries, Energy Environ. Sci. 5 (2012) 5163.

12. K. Kanehori, K. Matsumoto, K. Miyauchi, T. Kudo, Thin film solid electrolyte and its application to secondary lithium cell, Solid State Ionics 9–10 (1983) 1445.

13. N. Kuwata, J. Kawamura, K. Toribami, T. Hattori, N. Sata, Thin-film lithium-ion battery with amorphous solid electrolyte fabricated by pulsed laser deposition, Electrochem. Commun. 6 (2004) 417.

14. Y.-S. Park, S.-H. Lee, B.-I. Lee, S.-K. Joo, All-solid-state lithium thin-film rechargeable battery with lithium manganese oxide, Electrochem. Solid-State Lett. 2 (1999) 58.

15. N. Kuwata, N. Iwagami, Y. Tanji, Y. Matsuda, J. Kawamura, Characterization of thin- film lithium batteries with stable thin-film Li3PO4 solid electrolytes fabricated by ArF excimer laser deposition, J. Electrochem. Soc. 157 (2010) A521.

16. J.B. Bates, N.J. Dudney, B. Neudecker, A. Ueda, C.D. Evans, Thin-film lithium and lithium-ion batteries, Solid State Ionics 135 (2000) 33.

17. J. Kawamura, N. Kuwata, K. Toribami, N. Sata, O. Kamishima, T. Hattori, Preparation of amorphous lithium ion conductor thin films by pulsed laser deposition, Solid State Ionics 175 (2004) 273.

18. Y.-N. Zhou, M.-Z. Xue, Z.-W. Fu, Nanostructured thin film electrodes for lithium storage and all-solid-state thin-film lithium batteries, J. Power Sources 234 (2013) 310.

19. J. Hong, C. Wang, N.J. Dudney, M.J. Lance, Characterization and performance of LiFePO4 thin-film cathodes prepared with radio-frequency magnetron-sputter deposition, J. Electrochem. Soc. 154 (2007) A805.

20. K.-F. Chui, Optimization of synthesis process for carbon-mixed LiFePO4 composite thin-film cathodes deposited by bias sputtering, J. Electrochem. Soc. 154 (2007) A129.

21. J. Xie, N. Imanishi, T. Zhang, A. Hirano, Y. Takeda, O. Yamamoto, Li-ion diffusion kinetics in LiFePO4 thin film prepared by radio frequency magnetron sputtering, Electrochim. Acta 54 (2009) 4631.

22. M. Köhler, F. Berkemeier, T. Gallasch, G. Schmitz, Lithium diffusion in sputterdeposited lithium iron phosphate thin-films, J. Power Sources 236 (2013) 61.

23. C. Yada, Y. Iriyama, S.-K. Jeong, T. Abe, M. Inaba, Z. Ogumi, Electrochemical properties of LiFePO4 thin films prepared by pulsed laser deposition, J. Power Sources 146 (2005) 559.

24. F. Sauvage, E. Baudrin, L. Gengembre, J.-M. Tarascon, Effect of texture on the electrochemical properties of LiFePO4 thin films, Solid State Ionics 176 (2005) 1869.

25. J. Sun, K. Tang, X. Yu, H. Li, X. Huang, Needle-like LiFePO4 thin films prepared by an off-axis pulsed laser deposition technique, Thin Solid Films 517 (2009) 2618. D. Fujimoto et al. / Thin Solid Films 579 (2015) 81–88 87

26. S.-W. Song, R.P. Reade, R. Kostecki, K.A. Striebel, and Electrochemical studies of the LiFePO4 thin films prepared with pulsed laser deposition, J. Electrochem. Soc. 153 (2006) A12.

27. T. Matsumura, N. Imanishi, A. Hirano, N. Sonoyama, Y. Takeda, Electrochemical performances for preferred oriented PLD thin-film electrodes of LiNi0.8Co0.2O2, LiFePO4 and LiMn2O4, Solid State Ionics 179 (2008) (2011–2015).

28. Z.G. Lu, M.F. Lo, C.Y. Chung, Pulse laser deposition and electrochemical characterization of LiFePO4–C composite thin films, J. Phys. Chem. C 112 (2008) 7069.

29. K. Tang, X. Yu, J. Sun, H. Li, X. Huang, Kinetic analysis on LiFePO4 thin films by CV, GITT, and EIS, Electrochim. Acta 56 (2011) 4869.

30. Z.G. Lu, H. Cheng, M.F. Lo, C.Y. Chung, Pulsed laser deposition and electrochemical characterization of LiFePO4–Ag composite thin films, Adv. Funct. Mater. 17 (2007) 3885.

31. J. Ma, Q.-Z. Qin, Electrochemical performance of nanocrystalline LiMPO4 thin-films prepared by electrostatic spray deposition, J. Power Sources 148 (2005) 66.

32. W.C. West, J.F. Whitacre, B.V. Ratnakumar, Radio frequency magnetronsputtered LiCoPO4 cathodes for 4.8 V thin-film batteries, J. Electrochem. Soc. 150 (2003) A1660.

33. J. Xie, N. Imanishi, T. Zhang, A. Hirano, Y. Takeda, O. Yamamoto, Li-ion diffusion kinetics in LiCoPO4 thin films deposited on NASICON-type glass ceramic electrolytes by magnetron sputtering, J. Power Sources 192 (2009) 689.

34. A. Eftekhari, Surface modification of thin-film based LiCoPO4 5 V cathode with metal oxide, J. Electrochem. Soc. 151 (2004) A1456.

35. J.L. Shui, Y. Yu, X.F. Yang, C.H. Chen, LiCoPO4-based ternary composite thin-film electrode for lithium secondary battery, Electrochem. Commun. 8 (2006) 1087.

36. M.S. Bhuwaneswari, L. Dimesso, W. Jaegermann, Preparation of LiCoPO4 powders and films via sol–gel, J. Sol-Gel. Sci. Technol. 56 (2010) 320.

37. N. Kuwata, R. Kumar, K. Toribami, T. Suzuki, T. Hattori, J. Kawamura, Thin film lithium ion batteries prepared only by pulsed laser deposition, Solid State Ionics 177 (2006) 2827.

38. K.P. Korona, J. Papierska, M. Kaminska, A. Witowski, M. Michalska, L. Lipinska, Raman measurements of temperature dependencies of phonons in LiMnPO4, Mater. Chem. Phys. 127 (2011) 391.

39. Z. Weixin, R. Xiangbin, Y. Zeheng, W. Hua, W. Qiang, H. Fei, Hydrothermal synthesis of crystalline α/β-MnO nanorods via γ-MnOOH nanorod precursors, Front. Chem. Eng. China 1 (2007) 365.

40. D. Wang, C. Ouyang, T. Drézen, I. Exnar, A. Kay, N.-H. Kwon, P. Gouerec, J.H. Miners, M. Wang, M. Grätzel, Improving the electrochemical activity of LiMnPO4 via Mn-site substitution, J. Electrochem. Soc. 157 (2010) A225.

41. Craig A.J. Fisher, Veluz M. Hart Prieto, M. Saiful Islam, Lithium battery materials LiMPO4 (M = Mn, Fe, Co, and Ni): insights into defect association, transport mechanisms, and doping behavior, Chem. Mater. 20 (2008) 5907.

42. N. Kuwata, S. Kudo, Y. Matsuda, J. Kawamura, Fabrication of thin-film lithium batteries with 5-V-class LiCoMnO4 cathodes, Solid State Ionics 262 (2014) 165.

43. X. Yu, J.B. Bates, G.E. Jellison Jr., F.X. Hart, A stable thin-film lithium electrolyte: lithium phosphorus oxynitride, J. Electrochem. Soc. 144 (1997) 524.

44. J.E.B. Randles, A cathode ray polarograph. Part II.—the current–voltage curves, Trans. Faraday Soc. 44 (1948) 327.

45. F.G. Lether, P.R. Wenston, An algorithm for the numerical evaluation of the reversible Randles–Sevcik function, Comput. Chem. 11 (1987) 179.

CITATION

Daichi Fujimoto, Naoaki Kuwata, Yasutaka Matsuda, Junichi Kawamura, Feiyu Kang, Fabrication of solid-state thin-film batteries using LiMnPO4 thin films deposited by pulsed laser deposition, Thin Solid Films, Volume 579, 31 March 2015, Pages 81-88, ISSN 0040-6090, http://dx.doi.org/10.1016/j.tsf.2015.02.041.

CHAPTER 5

Dynamics of Continuously Pumped Solid-State Regenerative Amplifiers

Mikhail Grishin and Andrejus Michailovas

Institute of Physics & EKSPLA uab, Lithuania

INTRODUCTION

Regenerative amplifiers are extensively used for amplifying pulses generated by mode-locked oscillators (Koechner, 2006). This is a powerful technique providing several orders of magnitude gain, virtually unlimited by amplified spontaneous emission, (the well known nemesis for multi-pass amplifiers) (Forget et al., 2002). Such uniquely high gain is achieved due to multiple passes of optical pulse through the gain medium. Multiple passes are organized by placing the gain media in an optical resonator. The number of round trips is typically controlled by an electro-optic switch (Nickel et al., 2005) [occasionally with acousto-optic modulator (Norris, 1992)]. The optical switch also provides quality control of the optical cavity, suppressing lasing and reducing the time period when parasitic amplification of spontaneous emission takes place. In addition, the optical cavity provides "filtering" of spatial mode (there is no spatial imperfection accumulation during multiple passes of amplifying pulse). Consequently, the possibility exists to obtain perfect beam quality. At the same time the stable resonator does not allow expanding mode diameter too much (Magni, 1986), restricting capabilities for amplifying optical pulses to very high intensity - when it is required this job is placed to subsequent high aperture amplifiers (Siebold et al., 2008). An auxiliary technique, vitally important to amplify femtosecond pulses to high energies, is chirped-pulse amplification (the stretcher-compressor scheme) (Strickland & Mourou, 1985, Mourou & Umstadter, 1992).

In respect to the gain, the regenerative amplifiers probably have only one competitor – fiber amplifiers (Fermann et al., 2002, Liu H. et al., 2008). These amplifiers as well as all lasers based on optical fibers exhibit impressive progress over the last decade (Jeong et al., 2004). However, the extremely small mode diameter and large medium length intrinsic for optical fibers lead to significant influence of detrimental nonlinear effects. These distort the amplifying signal's optical spectrum eventually resulting in serious limitation of peak power so that 0.7 MW before compression is one of the best achievements (Röser et al., 2007). Regenerative amplifiers are commonly designed with bulk materials allowing larger mode diameter that moves away the peak power limit well above tens of megawatts (Kleinbauer et al., 2008).

Most frequently used material for amplification of femtosecond pulses (and thoroughly dominating below 100 fs) is titanium doped sapphire. Broad gain bandwidth, exceptionally good lasing properties and opto-mechanical characteristics place this medium in such a special position. High pulse energy and high average output power have been achieved by using Ti:saphire amplifiers (Walker et al., 1999,Matsushima et al., 2006). The disadvantages of this gain medium are related to possible means of pumping. First, corresponding absorption lines are located in the green spectrum range, where suitable laser diodes with reasonable power are not available. Second, the short gain relaxation time (3.2 µs) requires pulsed pumping with high pulse energy (usually with Q-switched frequency-doubled neodymium lasers) in order to store substantial population inversion and consequently to obtain high output energy. We should remark here that the issues described in present chapter are not valid for Ti:saphire regenerative amplifiers since we use approximations which suitable only for long lifetime media. From the other hand due to very short lifetime the Ti:saphire amplifiers do not exhibit extraordinary dynamic properties on which we are mainly focusing in the chapter.

Another family of popular laser materials is ytterbium doped crystals and glasses. Their wide spectrum supports amplification of ultrashort pulses (however not as short as Ti:sapphire supports). Moreover ytterbium materials allow direct laser diode pumping with intrinsically small quantum defect (typically 10% or even less). The latter enhances overall power efficiency and reduce heat generation in active elements, that in turn alleviates thermal effects which inhibit average power increase. Long upper-level lifetime of ytterbium ions virtually in all the crystals and glasses allows good capacity of stored energy under convenient continuous laser diode pumping. What somewhat challenges operation with ytterbium materials is relatively low gain (typically less than 10% per pass) which originated from peculiar to this materials small stimulated emission cross section. From the other hand regenerative amplification is indeed a way of efficient energy extraction at low gain; just special attention should be given to reduce inrtracavity losses (Biswal et. al., 1998). An alternate way to improve stimulated emission features is cryogenically cooled active elements (Kawanaka et al., 2003) although this method is bulky and, as a rule, it narrows gain bandwidth. Then special attention should be placed to host crystal selection not to limit development towards shorter pulses (Pugžlys et al., 2009).

Ytterbium doped media are able to withstand very intense optical pumping without detriment to exited-state population which in other materials can be limited by quenching effects (e.g. exited state absorption or up-conversion). This favorable property permits use of active elements in thin disc geometry. In particular, Yb:YAG thin disk lasers are scalable to very high average power and to high pulse energies (Speiser & Giesen, 2008). Extremely short optical pass within the thin disc reduces nonlinear effects (in essence the Kerr effect) allowing high peak power pulses even without using stretcher-compressor technique (Kleinbauer et al., 2008).

The amplifiers based on neodymium gain media have their specific advantages. High stimulated emission cross section simplifies system design and reduces requirements for optical components. Good energy storage capabilities allow high output energy and power efficient operation when pumping with laser diodes. Neodymium laser materials are well suited for picosecond pulse durations and are competitive for moderate average power. Systems based on Nd:YVO$_4$ and Nd:GdVO$_4$ crystals routinely produce more than 10 W of output power (Kleinbauer et al., 2005, Clubley et al., 2008).

The regenerative amplifiers of solid state lasers designed for scientific applications usually operate at low or moderate repetition rates (not exceeding several kHz). Presently, there is rising demand for high repetition rate ultrashort-pulse solid state lasers for material micro-processing (Meijera et al., 2002). On the other hand, a new generation of fast electro-optic switches became available such as Pockels cells based on -barium borate along with improved high-voltage electronics (Nickel et al., 2005, Siebold et al., 2004). As a result, picosecond and femtosecond lasers with repetition rate of the order of 100's kHz have come onto the market (Raciukaitis et al., 2006). Regenerative amplifiers are an important part of most ultrafast industrial solid state laser systems. Both high system efficiency and stable output parameters over a wide range of pulse repetition rates are essential for this actively developing field. For creation of power-efficient systems, neodymium and ytterbium laser gain media which may be directly pumped by laser diodes is advantageous. Long lifetime of the upper laser level typical of both these ions supports accumulation of substantial population inversion under continuous laser diode pumping. However, this long inversion lifetime may also cause stability problems at high repetition rates. Continuously pumped regenerative amplifiers demonstrate peculiar pulse amplification dynamics when the pulse repetition period becomes comparable or shorter than the gain relaxation time (Müller et al., 2003). Period doubling bifurcations develop generating periodically alternating energy pulses or even sequences of pulses having chaotic energy distribution.

Complex dynamic behavior is well known phenomenon in laser physics (Haken, 1975). Generally, nonlinear differential equations describing laser dynamics tend to have unstable solutions containing multi-stabilities and bifurcations when a number of independent variables representing system states are equal or more than three (Lorenz, 1963). By no means complete list of laser systems exhibiting complicated dynamics includes Q-switched gas lasers (Arecchi et al., 1982) passively Q-switched solid state lasers (Tang et al., 2003), optically injected solid state lasers (Valling et al., 2005). The specifics of high repetition rate regenerative amplifiers is such that their operation can be described with two differential equations, but periodic disturbance caused by release of the amplified pulse complicates the system behavior. Unlike many lasers which have been created specially to study dynamic phenomena and chaotic behavior, dynamics of regenerative amplifiers needs to be understood from a more utilitarian position in order to comprehensively optimize real systems. To date, only a few articles have been dedicated to this phenomenon despite its critical influence on the performance of regenerative amplifiers. Period doubling regime passing to chaotic operation has been observed for a system based on ytterbium doped glass and the role of bifurcations has been investigated both theoretically and experimentally (Dörring et al., 2004). However, one of the important parameters, the seed pulse energy, was left beyond the scope, and so applicability of the obtained results was restricted. The experiments were confined to studying cavity dumping of the Q-switched laser, an approximately equivalent system to the regenerative amplifier seeded by extremely low pulse energy. Our recent theoretical work has presented a generalized picture of stability features of a continuously pumped high repetition rate regenerative amplifier based on laser media with long relaxation time. The regions exhibiting different system behavior have been mapped in the space of non-dimensional control parameters: repetition rate and round trip number (Grishin et al., 2007). Additionally this analysis revealed the importance of the seed pulse energy and demonstrated that increase in the seed energy helps in eliminating the instabilities. Comprehensive utilization of these theoretical results has promoted top performance obtained from multi-kilohertz Yb:YAG disk amplifier (Metzger et al., 2009). Experimental study, performed with Nd:YVO$_4$ system, has thoroughly confirmed theoretical conclusions and on its basis the concept of regenerative amplifiers optimization has been formulated (Grishin et al., 2008, Grishin et al., 2009).

In the present chapter we summarize information from our already published papers and expand this description on the basis of our recent results. In the first theoretical part of the chapter we present a concept for modeling of continuously pumped solid state regenerative amplifiers. Simplified rate equations in normalized form allow reducing the system analysis to a task which has been thoroughly accomplished in theory of discrete-time dynamical systems (Alligood et al., 1996). Then we focus more attention to the influence of the main governing parameters on dynamics of regenerative amplification and limitations of the practical system performances occurred due to instabilities. We show why seed pulse energy plays such an important role at repetitive regime in contrast to low repetition rates. Influence of parasitic intracavity losses is shown to be a factor which not only decreases energy extraction efficiency (as in all laser systems) but also enhances instability range. We analyze numerically obtained stability diagrams which allow determination of optimal operation regime when maximum output energy can be extracted from the amplifier as stable pulse train. Important laser parameters which influence the system performance indirectly will be considered too. Heating of intracavity optical components depends not only on the range of output average power and the optical components quality but also on the amplification regime. The Kerr nonlinearity which limits system performance for short optical pulses will be evaluated taking into account multiple passes.

The verification of the created model is presented in the experimental part of the chapter. We investigate operation peculiarities of the system consisting of mode-locked master oscillator, the preamplifier and the regenerative amplifier based on Nd:YVO$_4$ crystals and analyze criteria of stable operation limitation. The experimental dependences of the output energy versus round trip number and repetition rate well agree with theoretical data. We demonstrate that increase of the seed pulse energy up to the value predicted by numerical model ensures stable operation within full range of repetition rates. The chapter is concluded by summarizing of presented theoretical and experimental results and suggesting of further investigations.

THEORY OF REGENERATIVE AMPLIFICATION IN REPETITIVE REGIME

Principle of Operation

Regenerative amplifier can be regarded as a system in which an optical resonator provides multiple passes, a gain medium is responsible for amplification, and an electro-optic switch serves as a valve in-turn admitting weak input pulse and releasing amplified pulse. At that, spatial properties of the amplified radiation are primarily determined by the optical resonator; the output energy mostly depends on the population inversion stored in the gain medium whereas the amplified pulse duration is imposed by the input seed pulse. A schematic diagram of a conventional solid-state diode-pumped regenerative amplifier is depicted in Fig. 1. This optical layout by no means differs from that for a common Q-switched laser and consequently there are certain similarities in operation (Murray & Lowdermilk, 1980). The resonator quality (Q-factor) is controlled by an electro-optic switch. The electro-optic switch usually consists of a Pockels cell, a quarter-wave plate and a polarizer. An operational cycle of the regenerative amplifier consists of two successive stages: low-Q and high-Q. When voltage is not applied to the Pockels cell the wave plate along with the polarizer provides high intracavity losses (low-Q state of the resonator). Laser action is suppressed by high losses, and the gain medium, being under continuous pumping, accumulates population inversion. The amplification takes place during high-Q stage. It starts when quarter-wave voltage is applied to the Pockels cell and the seed pulse is injected into the resonator. The intracavity losses become minimal and are kept low for some pre-set time while the optical pulse circulating in the resonator is amplifying, simultaneously consuming a certain part of the stored energy. The intracavity pulse energy grows until the gain becomes equal to the resonator losses and then the pulse energy decays. As soon as the intracavity energy reaches a desired level the Pockels cell voltage is switched off. This dumps the amplified pulse out of the cavity as the output pulse. The system returns into the initial state. Then during the next low-Q stage the depleted part of the inversion population is restored by uninterrupted pumping and the cycles iterate.

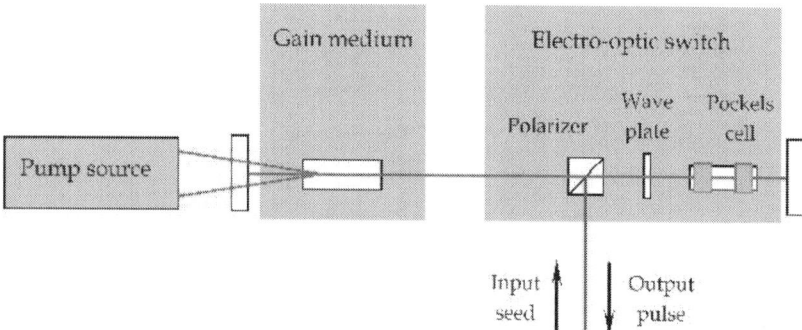

Figure 1. Optical layout of typical solid-state diode pumped regenerative amplifier.

At higher repetition rates, when the pump stage is comparable or shorter than the gain relaxation time, the operation cycles become interdependent. The equilibrium between population inversion depletion caused by amplification and inversion restoration caused by pumping may become unstable. This often leads to breakdown of the single-energy regime and to generation of periodically alternating high/low energy pulses, or more complicated instability patterns. We will consider this and related phenomena theoretically in the first part of the present chapter.

Rate Equations and Basic Terms

The process of regenerative amplification is essentially determined by the interaction of the intracavity radiation with the laser medium excited by pumping. The rules of this interaction can be established by using a simple phenomenological notion of a dynamic balance between basic processes: pumping and relaxation, amplification and extinction. Thus the system evolution can be described by coupled differential equations for the population inversion density N and photon number. We use the space independent rate equations which have been formulated for idealized four-level laser medium with a homogenously broadened line (Svelto, 1998):

$$dNdt = Rp - \sigma cV\varphi N - NT1 \tag{1}$$

$$d\varphi dt = (\sigma cLaAaVN - 1Tc)\varphi \tag{2}$$

The density of population inversion grows due to pumping (proportionally to the pumping rate Rp) and decays with the constant rate inversely proportional to the upper laser level lifetime T_1. This balance is significantly influenced by stimulated emission, the contribution of which depends on the gain medium characteristics and particularly on the stimulated emission cross section . The second equation establishes a rule of stimulated photons multiplication. The photon number increase proportionally to the population inversion and at the same time it decreases because of intracavity losses. The losses here are defined in terms of the photon lifetime Tc which determines the rate of decay for the light field in the optical resonator. Geometry of the amplifier is accounted for in terms of the mode volume within the optical resonator (V), the active medium length (L a) and the beam area in the active medium (A a). The velocity of light c is a constant coupling temporal and spatial terms of the equations. Since we use the rate equations approximation for modeling, the range of proposed model applicability is limited by the range of the particular rate equation's validity. Concerning general validity of the rate equation approach, we refer to (Svelto, 1998) for details. Here we just mention that this is a conventional method of laser dynamics study. It gives sufficiently accurate results for most practical purposes and in particular for analysis of multi-pass and regenerative amplification (Murray & Lowdermilk, 1980, Lowdermilk & Murray, 1980).

We shall re-arrange the equations to a form containing macroscopic non-dimensional terms. This is a common way to reduce the number of control parameters in order to provide simplified picture of system behavior without losing information. A primary normalization coefficient is the term proportional to the pump level, $G_0 = R_p T_1 L a$. The physical meaning of G_0 is steady state gain coefficient per pass. Such gain is achieved at the equilibrium of population inversion density (dN/dt=0) obtained under constant pumping (dRp/dt=0), constant rate of spontaneous decay (T$_1$) and in absence of lasing (=0). Thus we can proceed to equations for the normalized gain g=NL a/G$_0$ and normalized energy =/(A a G$_0$):

$$dg(t\sim)dt\sim = -\varepsilon(t\sim)g(t\sim)+1-g(t\sim)\tau1 \qquad (3)$$

$$d\varepsilon(t\sim)dt\sim = \varepsilon(t\sim)[g(t\sim)-gt] \qquad (4)$$

Additionally, we have modified the time scale by introducing normalized time t$\sim$=tG$_0$/T$_0$, a product of the current number of cavity round trips and the steady state gain coefficient. Here the denominator T$_0$represents the round trip time. As we will see later, the term t$\sim$ assigns a natural time scale to the high-Q phase of operation. The governing parameter, which defines the amplification period (effective round trip number), will be introduced on this basis. The normalized relaxation time of the gain$_1$=T$_1$G$_0$/T$_0$ is essentially determined by the upper laser level lifetime T$_1$. A term accounting for parasitic losses of the optical resonator is represented by normalized threshold gain, $g t$=T$_0$/(TcG$_0$). This parameter can be also expressed in terms of loss coefficient 1 as $g t$=ln (1l)/G$_0$, by using common threshold criterion of gain and losses equality.

Note that equations formulated with those new variables do not contain pumping characteristics in explicit form. Parameter G$_0$, proportional to the pumping rate, is hidden in the composition of basic variables. The pump effect (as well as other control parameters effects) is easy to restore when applying modeling results by performing reciprocal transformation from normalized to real parameters. Moreover, further in the theoretical part of this chapter we often omit the words "effective" and "normalized" just for shortening. Although Eqs. (3) and (4) have been obtained in the approximation of a four-level system, the identical equations can be formulated for the more general case of quasi-three-level systems [the initial rate equations and conditions of their applicability can be found in (Svelto, 1998)]. In the latter case the explicit expressions of normalized and effective terms look slightly more complicated but their physical meaning remains unchanged.

Consideration of a Single Operation Cycle

Below in this section we define relations between basic system parameters deriving the rate equations for separate stages of regenerative amplifier operation. First we consider the low-Q phase. Since during this phase of operation amplification is suppressed while pumping takes place, it is also called pump stage. Inasmuch as there is no lasing during this stage (i.e. =0), a set of Eqs. (3) and (4) transforms into a single equation:

$$dg(t\sim)dt\sim = 1 - g(t\sim)\tau 1 \qquad (5)$$

Initial conditions specify the gain at the beginning of the pump phase, g_{pi}. Taking into account that $t^~/_1=t/T_1$, we can find the relation between the initial gain g_{pi} and the final gain g_{pf} for a certain pump phase duration T:

$$gpf=1-(1-gpi)exp(-TT1) \qquad (6)$$

In the next section we will often use diminished form of the equation (6): $gpf=g^p(gpi,T/T1)$, where the function g^p just establishes a rule of the gain transformation $g_{pi} g_{pf}$ for the pump stage.

Then let us proceed to the high-Q phase. The equations for this operation stage can be simplified as well. We remind here that we explore laser media having long relaxation time, the case functionally important for diode-pumped systems. Since the buildup time of the optical pulse is usually short in comparison with the pump phase duration and the upper laser level lifetime, the population inversion change due to pumping and relaxation processes is much smaller than the inversion depletion caused by amplification. Hence, the terms containing spontaneous decay and optical pumping can be neglected in Eq. (3). Also, as we have assumed negligible pump contribution during high-Q phase, this stage of operation can be called in a more common and more informative manner as the amplification stage.

The next assumption presumes low intra-cavity losses ($g_t 0$). This approximation substantially simplifies the basics of the presented method. The simplification appears not only due to existence of the analytical solution for the rate equations but also, more importantly, because of the reduced number of parameters governing the system. An influence of the parasitic losses on amplification dynamics will be accounted in section 2.7 after the essence of the theoretical approach has been presented. For the present, we come to the situation at which Eqs. (3) and (4), when describing amplification stage, reduce to the following:

$$dg(t~)dt~=-\varepsilon(t~)g(t~) \qquad (7)$$

$$d\varepsilon(t\sim)dt\sim=\varepsilon(t\sim)g(t\sim) \tag{8}$$

Initial conditions specify the system state at the beginning of the amplification phase: the initial gain gai and the energy of the input pulse from which the amplification starts – the seed energy, s. Also it is natural to constrain our consideration to the case of low seed energy with respect to the stored energy ($s<<g\,ai$ in terms of normalized parameters). As a consequence of these assumptions, the solutions of coupled Eqs. (7) and (8) can be found in analytic form. Such solutions obviously describe temporal evolution of the gain and intracavity energy. One can also find other physical sense, more convenient for further consideration. In case of fixed amplification phase duration, temporal evolution is terminated at the moment $t\bar{}=$. Then the solutions also express how the system parameters on the amplification phase completion (final gain $g\,af$ and output pulse energy f) depend on the initial conditions and governing parameters:

$$g(t\sim)=gai1+\varepsilon sgaiexp(gait\sim)\Rightarrow gaf=gai1+\varepsilon sgaiexp(gai\tau) \tag{9}$$

$$\varepsilon(t\sim)=gai1+gai\varepsilon sexp(-gait\sim)\Rightarrow \varepsilon f=gai1+gai\varepsilon sexp(-gai\tau) \tag{10}$$

Here the terms and s are actual parameters controlling the amplification, whereas the initial gain is a variable coupling the equations. The normalized amplification stage duration is a product of cavity round trip number and steady state gain G_0. Further we will call this term in a more comprehensible manner an "effective round trip number". Essentially, the term also represents total multi-pass small signal gain factor for the amplifier. The Equations (9) and (10) can be presented in diminished form:

$gaf=g\hat{}a(gai,\varepsilon s,\tau)$ and $\varepsilon f=\varepsilon\hat{}(gai,\varepsilon s,\tau)$, simplifying further operation with these formulas. Now, as the basic terms have been introduced, we can summarize a rule for subscript notations. The pump and amplification operation phases are designated with p and a subscripts respectively; the initial and final states are notated with i and f subscripts correspondingly.

Coupling of Successive Cycles. Discrete-Time Dynamical System Approach

Evaluation of the output energy is a trivial task for low repetition rates, i.e. when the pump phase duration significantly exceeds the inversion relaxation time, $T \gg T_1$. The gain in this case reaches saturation before the amplification phase starts, that is the initial gain g_{ai} tends to unity. Consequently, the output energy versus round trip number is strictly defined by Eq. (10) alone. In general case, and particularly at high pulse repetition rate the initial conditions for the current operation stage depend on the previous system state. Hence, the initial gain for each cycle depends not only on operation parameters but also on the system pre-history. In order to determine the gain at the beginning of amplification we shall relate final and initial states of successive operation cycles.

Now we proceed from evaluation of the gain and pulse energy within single operation stages to a description of those stages as a whole by analyzing the temporal evolution of the boundary values. We assign initial and final gains and pulse energy as variables defining the system state. The term $g_{ai}(1)$ is introduced as the initial gain of the amplification phase for the first cycle of operation. This stage finishes with the final gain denoted as $g_{af}(1)$. The subsequent pump phase of the current cycle obviously begins from the same gain value, $g_{pi}(1)=g_{af}(1)$. Similarly, the gain evolution continuity should be taken into account for coupling of all operation cycles. There is a boundary relation $g_{af}(k)=g_{pi}(k)$ within the cycle number k and for subsequent cycles: $g_{pf}(k)=g_{ai}(k+1)$. The legend of the gain evolution in a discrete time scale can be presented as follows:

$$...g_{ai}(k) \rightarrow g_{af}(k)=g_{pi}(k) \rightarrow g_{pf}(k)=g_{ai}(k+1) \rightarrow g_{af}(k+1)=g_{pi}(k+1) \rightarrow g_{pf}(k+1)... \quad (11)$$

The corresponding time boundary points of neighboring operation stages can be described as $t_{ai}(k)=(k1)(T_0/G_0 T)$ and $t_{af}(k)=kT_0/G_0(k1)T$. Note that in an assumption of short amplification phase, the duration of complete cycle (dumping period) is approximately equal to the pump phase duration T. Hence the term $(T/T_1)^1$ represents normalized pulse repetition rate of the regenerative amplifier (also called dumping frequency).

Analogous transition to the discrete time scale can be applied to energy evolution. However, unlike continuity of the gain evolution, the build-up of intra-cavity energy $sf(k)$ interrupts at the end of the amplification phase of each cycle at the moment of pulse dumping and then it begins again from the constant level which corresponds to the seed pulse energy s. Hence the term $f(k)$, determining the output energy, does not depend of its own pre-history. It is dependent on the gain and can be found from Eq. (10) for any operation cycle provided that the gain is known. Therefore, the gain becomes the only independent variable that needs to be analyzed. We can consider the expressions obtained earlier $g\hat{}p$ for the pump phase and $g\hat{}a$ for the amplification phase explicitly represented by Eq. (6) and Eq. (9) as the rules of the system state updating. These solutions of the rate equations serve as transformation rules that take the current state as input and update it by producing a new state. This new output state becomes the input for the next stage of operation. Then it is possible to combine amplification and pump phases within certain operation cycle and to form a joint gain transformation rule. We introduce $g\hat{}\Sigma$ as the composition of functions $g\hat{}a$ and $g\hat{}p$ exhibiting the gain transformation rule for the complete cycle, $g\hat{}\Sigma=g\hat{}p(g\hat{}a)$. Then using expressions of the inner functions we can present an explicit form of the function $g\hat{}\Sigma$:

$$g^{\wedge}\Sigma(gai)=1-[1-gai1+\varepsilon sgaiexp(gai\tau)]exp(-TT1)$$

(12)

Thus, we have reduced the regenerative amplification to the evolution of single variable (system state, gai) in a discrete time scale; and also we have found a rule of this variable updating. The basic properties of this updating function fit to the mathematical definition of so called maps [functions whose domain (input) space and range (output) space are the same]. Then the regenerative amplification can be described by using the theory of one-dimensional discrete-time dynamical systems (one-dimensional maps) (Alligood et al., 1996). The sequence of the system states $g\,ai(1)$, $g\,ai(2)$,... $g\,ai(k)$ ... is called an orbit in terms of this theory. The orbits can be calculated by using a recurrent formula determining the subsequent state of the system in terms of the present state:

gai(k+1)=g^Σ[gai(k)]

It is obvious that in a regular single-energy regime the gain depletion during the amplification phase should be compensated by restoring the population inversion during the pump phase. In terms of states evolution, the initial gain of the amplification stage eventually should iterate, i.e. there is a certain gain value (designated as g_1) such that the subsequent gains stabilize upon reaching that value, $g\,ai(k1)=g\,ai(k)=g_1$. Consequently, the system eigenstate satisfying the condition g1=g^Σ(g1) should exist. The solution of this equation is known as a fixed point in the discrete-time dynamical system theory. Being exactly equal to the fixed point, the system state reproduces itself after each cycle that leads to operating in a regular manner. However, requirement of technical feasibility of such a regime establishes a more strict condition to be fulfilled. The system should return to the fixed point after some perturbation has occurred, in other terms, more common for theory of dynamical systems, the fixed point should be attracting. Thus, study of regenerative amplification is reduced to the analysis of conditions of the fixed point existence and its stability characterization.

Since the equation determining the fixed points is transcendental we start analysis of the system state evolution with the graphical illustration. For more intuitive presentation, the fixed point existence condition g1=g^Σ(g1) with the explicit form of g^Σ given by Eq. (12) is rearranged into the following form:

$$1-(1-g1)\exp(TT1)=g11+\varepsilon sg1\exp(g1\tau) \tag{13}$$

The right-hand part of this expression represents the gain transformation function during amplification, g^a [see Eq. (9)]. The left-hand part may be regarded as an inverse function of the gain recovery during the pump phase transformed Eq. (6) gives gpi=g^-1p(gpf). Since gain continuity implies equality of the boundary states (g_{ai}=gpf and gaf=gpi) we can combine both curves gaf versus gai and gpi versus gpf on a common diagram (Fig. 2). A space of system states, defined by this means, can give an intuitively simple but strict and fruitful picture of the system state evolution. The intersection of those curves, having clear physical meaning, gives solution of Eq. (13), i.e. it determines the fixed point of the system. It is important that the intersection of these curves always exists and it is always single for any set of control parameters. Really, the amplification stage curve is single-peaked, it begins from zero and always lies under the state space diagonal (gaf=gai). The letter is natural because during the amplification stage (provided, as we assumed, negligible pumping contribution) the population inversion is depleting by transforming to the intracavity pulse energy and respectively the gain can only decrease, gaf<gai. The pump stage curve is a straight line whose slope depends on the repetition rate. This curve begins in the right upper corner of the state space [(1, 1) point] and also always locates under the state space diagonal. Note, the state space, due to proper normalizing, has dimensions of (0-1)(0-1). Thus, these curves cannot help intersecting and they intersect only once. Moreover, since the basic properties of the curves are universal, a fixed point existence and uniqueness is not only the result of mathematical speculations obtained under certain approximations but also the consequence of inherent physical properties of regenerative amplification. One of two necessary requirements for existence of a stable single-energy regime, namely the fixed point uniqueness, is fulfilled and then the main concern is the fixed point stability study. Figures 2(a) and 2(b) represent diagrams of system state evolution for two typical cases. Figure 2(a)) presents the orbit converging into an attracting fixed point. Such a convergence means that the regenerative amplifier eventually (after sufficient number of reiterations, when initial value of the orbit is "forgotten") starts producing regular pulsing. It is intuitively seen that the behavior of the resulting orbit (convergent or non-convergent) depends on the slope of the "amplification" curve in the fixed point with respect to the slope of the "pump" curve. Strictly speaking, the fixed point becomes attracting if the derivative of g^Σ function in this point satisfies the requirement |g^Σ(g1)|1 (Alligood et al., 1996). The condition |g^Σ(g1)|=1 represents the transition point between stable and unstable operation. In

case of $|g\hat{}'\Sigma(g1)|1$ the fixed point is repelling, and consequently stable operation becomes unfeasible.

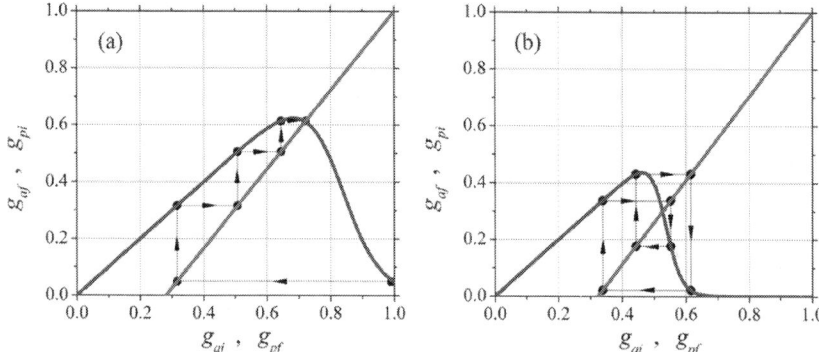

Figure 2. Graphical presentation of the orbits in state space. The fixed point is the intersection of the "amplification" and "pump" (blue and red) curves. Transition to the stable (attracting) fixed point (a) at $s=310^7$; $=18.0$; $T_1/T=3.0$. Period-4T orbit (b) at $s=10^{10}$; $=42.0$; $T_1/T=2.56$.

It becomes apparent that in the latter case the system is unable to reproduce its own state after one cycle of operation. However, such a iteration may occur after two or several cycles. The corresponding set of system states is called periodic orbit. An example of periodic orbit is depicted in Fig. 2(b). The condition for existence of the orbit with double period, 2T can be written by introducing an appropriate composite function. Define $g\hat{}2\Sigma=g\hat{}\Sigma(g\hat{}\Sigma)$ to be the result of applying the map-function $g\hat{}\Sigma$ to the system state two times. The system state g_2, such that $g2=g\hat{}2\Sigma(g2)$, is the fixed point analogue but suited for two successive operation cycles. Generally the orbit with the period of mT exists if there is a system eigenstate gm satisfying the equation: $gm=g\hat{}m\Sigma(gm)$. Here the term m is an integer number exhibiting a factor of output pulse repeatability for the corresponding multi-energy regime. If such a regime is realized, the system produces quasi-periodic sequence of the output pulses. The pulses of identical magnitude in this sequence appear each time in a multiplied period equaled to mT. In much the same way as the existence of a fixed point does not ensure stable operation, the existence of a periodic orbit does not in itself mean that the corresponding regime is realizable. Additional analysis of the orbit stability is required, that, similar to the fixed point case, reduces to evaluation of the corresponding map-function derivative. The orbit $gai(k)$ of period-m is stable provided that $|(g\hat{}m\Sigma)'(gm)|1$. Computation of the derivative for this composite function is feasible since its value eventually (at $k\infty$) tends to the product of its inner function derivatives at points along the orbit: $(g\hat{}m\Sigma)'(gm)=g\hat{}'\Sigma[gai(1)]g\hat{}'\Sigma[gai(2)]\cdots g\hat{}'\Sigma[gai(k)]$. If the absolute value of the product of the derivatives is larger than one, then periodicity of the orbit becomes unfeasible meaning that the system exhibits chaotic behavior. This is the Lyapunov number criterion of deterministic chaos (Alligood et al., 1996).

We can remind here that the map-function $g\hat{}\Sigma$ is itself a function of system parameters $(, s, T/T_1)$. As one of the governing parameters is varied the corresponding fixed point passes through different states of stability. A pass through the position $|(g\hat{}m\Sigma)'[gai(k)]|=1$ causes qualitative change of the system operation. Such transitions (e.g. transition from stable to unstable regime at m=1) is usually called a bifurcation. A set of control parameters (operation point) at which the bifurcation occurs is referred to as the bifurcation point. Correspondingly the diagram of the output parameter versus one of the control parameters for the system exhibiting bifurcations is often called a bifurcation diagram. Among many possible types of bifurcations, known for dynamical systems, we have met here the bifurcation of period doubling. This relatively simple type of dynamic behavior is one of the consequences of the fixed point uniqueness. This attribute gives also primary unambiguity of the system behavior. The dynamics of regenerative amplification and output characteristics of the system are determined by the set of control parameters alone in contrast to e.g. bi-stability effects where initial value of the orbit, $gai(1)$ may also influence the operation. One can imagine the latter as qualitative change of system behavior caused by a way of switching it on (e.g. in practice either one has the pumping source enabled first and then seeding or other way around). The unambiguous relation between control parameters and operation regimes is quite an important property of regenerative amplifiers and our further analysis always implies this property without necessarily mentioning it.

Diagrams of Dynamic Regimes in the Parameter Space

Simplifying assumptions and non-dimensional effective parameters, introduced for the basic rate equation model, reduce the number of independent control parameters of the system to the set of three. These are the normalized repetition rate, T_1/T, amplification phase duration expressed in terms of the effective round trip number and normalized seed energy s. Analysis of stability for orbits of the initial gain of amplification phase [$gai(k)$] at each given control parameter provides thorough picture of regenerative amplifier behavior. The orbits were calculated by iterating of Eq. (12) in the range of control parameters wide enough to comprehend all the relevant dynamics features: $0.2<(T_1/T)<20$; $10<<110$; $10^{11}<s<10^3$. We used as much as 3000 iterations, the sufficient number to be confident that the results are independent of the system state at the beginning of iterating. The orbits were analyzed in two stages. At first, the minimal number of cycles between repeating system states was revealed for each orbit in parameter space. It was performed by direct comparison of the system state sequences with themselves but shifted by a certain cycle number, $g_{ai}(k)$ versus $g_{ai}(k+m)$. In that way the periodic orbits up to $m=32$, including regular ones ($m=1$), were identified. Then the Lyapunov number criterion was applied to the residual unidentified orbits. They were separated into two fundamentally different bunches: chaotic and eventually periodic.

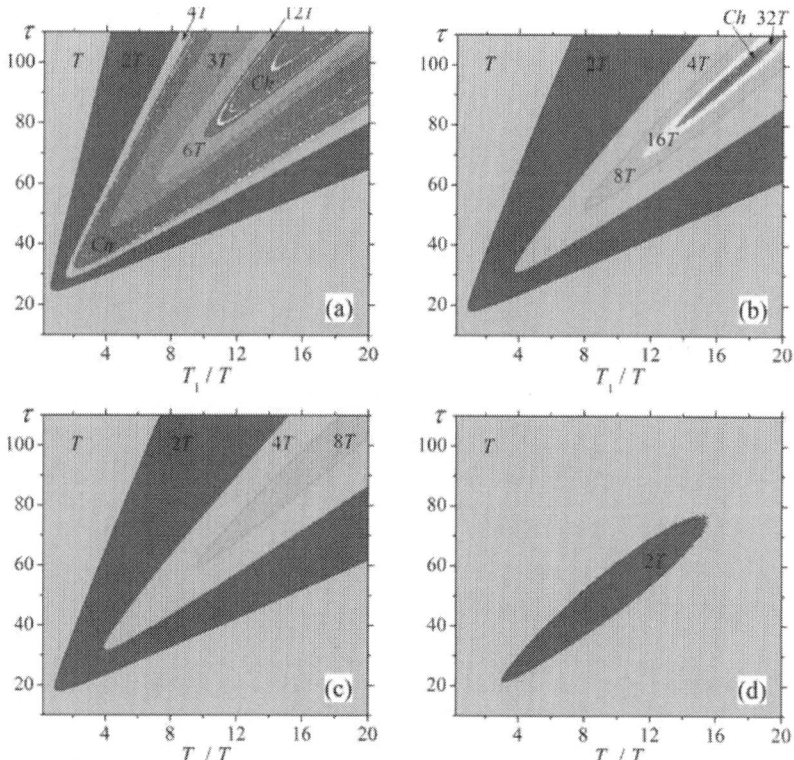

Figure 3. Diagrams of amplification dynamics in parameter space for different seed energies: $s=10^{10}$ (a); $s=2.5\,10^7$ (b); $s=3\,10^7$ (c); $s=1.3\,10^4$ (d).

Thus, the following dynamic regimes were distinguished in three-dimensional space of control parameters in accordance with the orbits properties: (1) The orbits evolving into stable fixed points ($m=1$)

corresponding to the regular system behavior (single pulse energy output, i.e. 1T-regime). (2) Periodic orbits corresponding to multi-energy regimes with repeatability coefficients in the range of $2 \leq m \leq 32$. (3) Eventually periodic orbits having larger repeatability factor ($m > 32$), for which the m-number itself is not identified. (4) Regime of deterministic chaos in accordance with the Lyapunov number criterion.

The regions of different dynamics are mapped in space of the repetition rate round trip number (Fig. 3). The major part of the parameter space is occupied by the regions corresponding to the following regimes: single-energy (1T); quasi-periodic with fundamental period of two (2T, 4T, 8T, 16T, and 32T); quasi-periodic with fundamental period of three (3T, 6T, and 12T); and chaotic behavior. These domains are marked with different colors, whereas the rest of the space containing the remaining zones of eventually periodic orbits is left white. The boundaries between adjacent colors (i.e. between different regimes) represent manifolds of bifurcation points in parameter space.

As it is seen, the dynamics turned out multifarious. Chaotic regime ordinarily comes out from the chain of successive period doubling bifurcations: T-2T-4T-8T-16T-32T... The chaotic zone itself has fine structure. Quasi-periodic "windows" with various periods are disseminated in it. The dynamics of regenerative amplification strongly depends on the seed value. The pattern is complex for low seed level ($s < 10^9$), the parameter space contains more than one clearly distinguishable chaotic regions [Fig. 3(a)]. Quasi-periodic regimes with fundamental period of three are observed between zones of chaotic dynamics. The higher the seed energy, the simpler the instability pattern becomes. Initially, chaotic domain shrinks to ellipse [Fig. 3(b)] and disappears from the parameter space. Furthermore, period doubling bifurcations with fundamental period of two only remain for $s > 2.5210^7$. Then the maximum order of bifurcations decreases [Figs. 3(c) and 3(d)] and finally, at $s > 1.910^4$ the system becomes stable in the whole range of control parameters.

Seed Pulse Energy Effect

The obtained results, demonstrating dependence of the operation on the seed pulse energy, are in essence not quite trivial. This phenomenon is in controversy with intuitive comprehension of regenerative amplification. The following speculations seem to demonstrate the negligible extent of the seed influence or at least to evidence much simpler looking relations. Imagine, initially low seed pulse energy s_1 after certain number of round trips () is amplified to energy s_2 of several orders of magnitude larger but still much less than energy stored in the gain media, $g_{ai} \gg s_2 \gg s_1$. Then further amplification should give the same output as if the amplification has initially started with seed energy s_2 because the previous stage ($s_1 s_2$) virtually has not changed the stored energy and, as a consequence, the system gain. This logic leads to an inference that a lack of seed energy can be compensated with additional round trips. Consequently, the operation diagrams have to look identical but shifted in the coordinate of round trip number for different seed values. Accurate computations give absolutely different results, Fig. 3.

Let us consider in details some subtleties which lead to this difference. In time-domain the amplification process looks exactly as simple logic predicts, that can be confirmed with straightforward calculations. Equation 10 gives equal but shifted in time intracavity energy evolutions, (g_{ai}, s_1, t)=(g_{ai}, s_2, t) and this shift, can be determined as: $g_{ai}\Delta\tau = \ln g_{ai}\varepsilon s_1 - \ln g_{ai}\varepsilon s_2$. This logic is accurate when we are considering amplification phase as "isolated" with given initial gain; this is absolutely true for low repetition rates.

Previously it was commonly accepted that regenerative amplification virtually is independent of the seed energy because only low repetition rates were under consideration. In reality, even for low repetition rate systems the seed energy value should not be too low. However, the reason for that is rather different of what we are describing. Simply, competing parasitic processes of amplification of spontaneous emission always exists in regenerative amplifiers. Thus, the seed energy should be well above the spontaneous emission level in order to get the amplified seed at the output instead of amplified spontaneous emission. However, at low repetition rates the sufficient pulse seed energy is extremely small, e.g. down to 10^{15} J in accordance to the experiments presented in the classical paper

(Murray & Lowdermilk, 1980). Although, some exceptions may take place; they involve special applications which require very clean, high contrast optical pulse, e.g. parametric chirped pulse amplification (Ross et al., 2007).

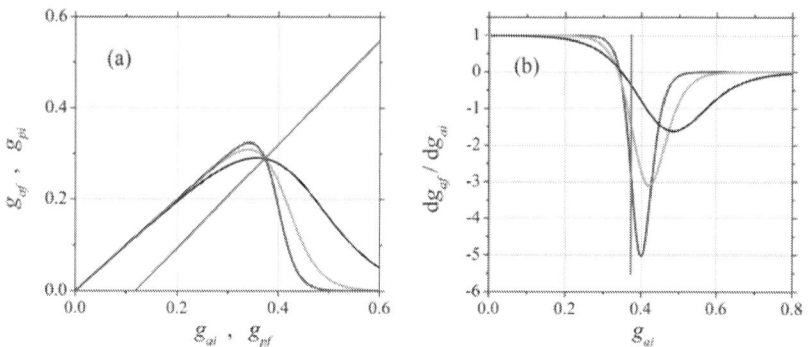

Figure 4. Typical state space diagrams of regimes having equal fixed points at different seed energies (a) and derivatives of the „amplification" curves (b).

It becomes apparent that at high repetition rates the initial gain depends on preceding operation cycles and can be determined indeed by taking them into account. This procedure is in essence nothing else than the fixed point determination. Let us return to geometric presentation of fixed points in state space. Figure 4(a) represents diagrams of the final gain against initial gain for three seed energy values. Corresponding numbers of round trips have been selected so that the fixed points at certain repetition rate are identical. All the curves intersect in a single point which corresponds to the fixed point for the repetition rate $T_1/T=8.0$ This means that decrease in seed energy is compensated by increasing of round trips but only in a sense of equal position of the fixed points. However the shapes of curves g_{af} versus g_{ai} are different and their derivatives in the intersection point are dependent on the seed energy [Fig. 4(b)]. Such derivatives indeed determine the regenerative amplification stability as we have described in section 2.4 by referring to the theory of discrete-time dynamical systems.

The identity of the fixed points can be realized by compensation of the seed energy difference by appropriate selection of the round trip number. Output energies corresponding to those fixed points are equal too. However the peculiarity of operation at high repetition rates is such that dynamical system behavior is absolutely different. Thus at high repetition rates the seed pulse energy becomes one of those critical parameters which determine the operation regime of the regenerative amplifier. The specific operation points, which have been analyzed here, can be also found in the diagrams of dynamic regimes (Fig. 3). With regard to stability, they were classified as stable, 2T-periodic and chaotic for seed values of 1.310^4, 2.510^7 and 10^{10} respectively. The dynamic regimes which are in general possible to obtain (by changing the round trip number) within a certain range of seed values are summarized in Table 1.

Table 1. Possible regimes versus seed pulse energy range.

Existing regimes	Seed value range	
	$g_t=0$	$g_t=0.028$
Chaos and "all" periods	$<2.52\ 10^{(7}$	$<1.4\ 10^5$
T , 2 T , 4 T , 8 T , 16 T , 32 T...	$2.52\ 10^{(7} - 2.56\ 10^{(7}$	$1.4\ 10^{(5} - 1.5\ 10^{(5}$
T , 2 T , 4 T , 8 T , 16 T	$2.56\ 10^{(7} - 2.72\ 10^{(7}$	$1.5\ 10^{(5} - 1.74\ 10^{(5}$
T , 2 T , 4 T , 8 T	$2.72\ 10^{(7} - 3.56\ 10^{(7}$	$1.74\ 10^{(5} - 2.5\ 10^{(5}$
T , 2 T , 4 T	$3.56\ 10^{(7} - 1.39\ 10^{(6}$	$2.5\ 10^{(5} - 4.1\ 10^{(5}$
T , 2 T	$1.39\ 10^{(6} - 1.90\ 10^{(4}$	$4.1\ 10^{(5} - 3.5\ 10^{(3}$
T (table)	$"/1.90\ 10^{(4}$	$"/3.5\ 10^{(3}$

Influence of Parasitic Intracavity Losses on Dynamic Pattern
The approximation of negligible losses is a good way to present the main ideas for application of the discrete-time dynamics method for regenerative amplification and to understand the dynamic patterns most relevant at high repetition rates. However, this approximation has limited application in practice. The output pulse energy grows in the lossless system monotonically together with amplification phase duration and reaches saturation at the level of $max=1\exp(T/T_1)$ that corresponds to full conversion of stored energy (population inversion) to the output pulse energy. Consequently the number of round trips can be increased, without detriment to output energy, to the values high enough for operation behind

the bifurcation zone that in turn assures stable operation. Actually, the system is always (i.e. irrespective of losses) stable provided that the population inversion is well depleted during the amplification phase. In this case the initial gain tends to the constant, determined only by the repetition rate, [$g_{af}0$ $g_{pi}0$; then from Eq. 6 follows that $g_{ai}=g_{pf}1\exp(T/T_1)$]. Consequently, the interdependence of operation cycles vanishes that results in eliminating of immediate cause of unstable behavior. In reality, parasitic losses prevent utilization of this property since because of losses the mode of complete gain depletion becomes inefficient.

Well known efficiency criterion, to dump optical pulse off the resonator at the moment when the current gain has dropped down to the threshold gain ($g_{af}=g_t$), is not applicable to repetitive operation as relating to only "isolated" operation cycles. Power efficiency enhancement takes place at high repetition rates when stored energy is left partially under-depleted ($g_{af}>>g_t$) forming a substantial gain background after several operation cycles. The proportion of gain to losses, which eventually determines extraction efficiency of the stored energy, can be well improved by this means. However incomplete depletion is an origin of operation cycles interdependence therefore in presence of losses the system efficiency in some sense collides with the system stability.

Parasitic losses in laser systems are given by optical components imperfection and diffraction losses of the optical resonator. The latter are objects of resonator geometry optimization. In case of solid-state lasers pumped longitudinally (virtual absence of hard apertures) high order optical aberrations (spherical e.g.) may become the main origin of diffraction losses, especially at high pumping intensities (Liu C. et al. 2008). Among optical components, the electro-optic switch is usually the most critical part; contributions of the Pockels cell and the polarizer to the loss factor surpass the remaining components (Müller et al., 2003). Practically, the level of total parasitic losses can vary in quite a wide range, but the typical value should not exceed a few percent per roundtrip for high-quality, well optimized systems. Here we should remark that the losses, inherent for quasi three-level gain media and related to partially populated ground state, are not dissipative and they do not belong to the parasitic losses which we are considering.

At the account of intracavity losses, general, qualitative pattern of amplification dynamics (fixed points uniqueness, variety of orbits for the repulsive fixed points, significance of the seed pulse energy) remains the same, but naturally the quantitative difference factors in. The intracavity losses of regenerative amplifier not only reduce efficiency (that is natural for lasers), but also substantially interfere in total system stability.

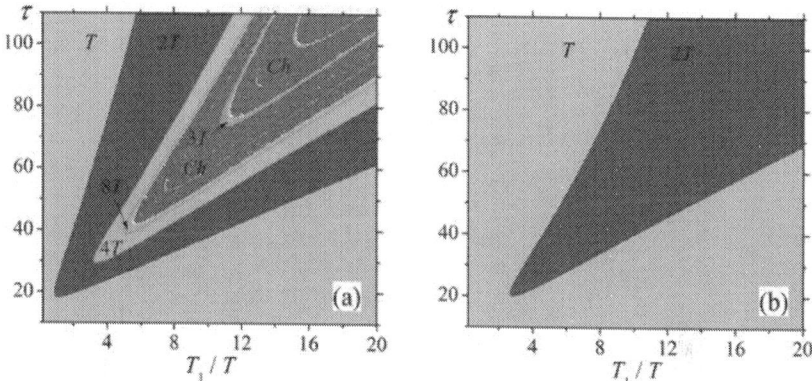

Figure 5. Diagrams of amplification dynamics for threshold gain $gt=0.028$ and seed pulse energies $s=3\,10^7$(a); $s=1.9010^4$ (b).

The diagrams of regenerative amplification dynamics in parameter space (round trips – repetition rate) for intracavity losses corresponding to the threshold gain $gt=0.028$ are presented in Fig. 5. Fixed point calculation and their stability evaluation were performed analogously as described in sections 2.4 and 2.5. The only difference is function $g\hat{}a$, relating final and initial gains [solutions of Eqs. (4) and (7)], was calculated numerically since analytical solution is unknown in case of nonzero losses. The diagram presented in Fig 5(a) can be compared with that given for the same seed pulse energy $s=3\,10^7$ but for zero losses [Fig 4(c)]. The influence of losses results in a more complicated dynamic pattern; the zone of chaotic dynamics evolves and the high order bifurcations shift closer to the tip of the instability zone (towards lower repetition rate and lower round trip number). The second diagram [Fig 5(b)] was calculated for the seed energy $s=1.9010^7$, which at zero losses provided stable operation in the whole range of control parameters. Now a period doubling zone (2T) has occupied a certain part of the parameter space and noticeably narrowed the range of stable operation.

A more cumulative picture of dynamical regimes is presented in Table 1. One can see some general change for the worse for system stability with respect to the zero-loss case. The decrease in stability caused by losses is not an obvious phenomenon (why not increase?). The reason for that bears similarity to the seed energy effect. The losses decrease pulse energy addition per round trip that can be compensated by increasing of round trip number, but only in a sense of fixed point identity. The derivative magnitude of the gain transformation function, $|g^{\wedge}\Sigma(g1)|$, in this point has changed so that system stability becomes worse as the losses are increasing. This phenomenon is not obvious but the conclusion is straightforward – the parasitic losses should be minimized as much as technically possible not only for efficiency but also for better stability.

System Performance Representation in Parameter Space

For understanding regenerative amplifier performance, not only dynamical regimes but also output pulse energy should be represented in relation to the parameters governing the system. In previous sections we have described in detail the method of fixed points and related orbits determination. Actually, during calculation of the initial gain orbit, $g\,ai\,(k)$, the orbit of output energy $f(k)$ comes out automatically as the second solution of coupled Eqs. (4) and (7), $\varepsilon f(k)=\varepsilon^{\wedge}[gai(k),\varepsilon s,\tau,gt]$. If governing parameters are such that operation is unstable then the initial gain varies from cycle to cycle by specific means and consequently the output energy becomes a multi-valued function of governing parameters. Representation of such a function on 2D diagram is unfeasible, it looks like a mess. Therefore we present typical dependencies of output energy versus round trip number (bifurcation diagrams) at several repetition rates for the fixed loss factor (corresponding to the threshold gain $gt=0.028$) and seed energy $s=7.710^7$ (Fig. 6). The selected seed pulse energy belongs to the same range that typical for functionally important class of seed lasers operating in CW mode-locking regime with moderate average power (around hundreds milliwatt). The repetition rates are chosen according to the diagram of dynamic regimes [Fig. 6(a)] so that typical bifurcation diagrams up to 16T-regime are demonstrated.

It is seen that the output energy variation in the presence of period doubling is so high that it virtually leaves no opportunity to use such a regime in practical applications. For example at repetition rate $T_1/T=4.6$ and in 30-60 round trips range the output energy alternates between high and low value so badly that the output pulse train looks almost as at twice less repetition rate [Fig. 6(c)]. It is even more pity that such bad stability often appears in regimes which potentially capable of providing high output energies.

The maximum capability of the system can be determined by calculating the output energy exactly in the fixed point: $g\,ai=g_1\ \varepsilon 1=\varepsilon\hat{\ }(g1,\varepsilon s,\tau,gt)$. This energy is always a single-valued function of the governing parameters regardless of whether the fixed point is attractive or repulsive. In case of a repulsive fixed point (i.e. unstable operation), the corresponding "fixed point" output energy becomes an artificial parameter but it can serve as a convenient reference for evaluating the power efficiency reduction caused by instability effects.

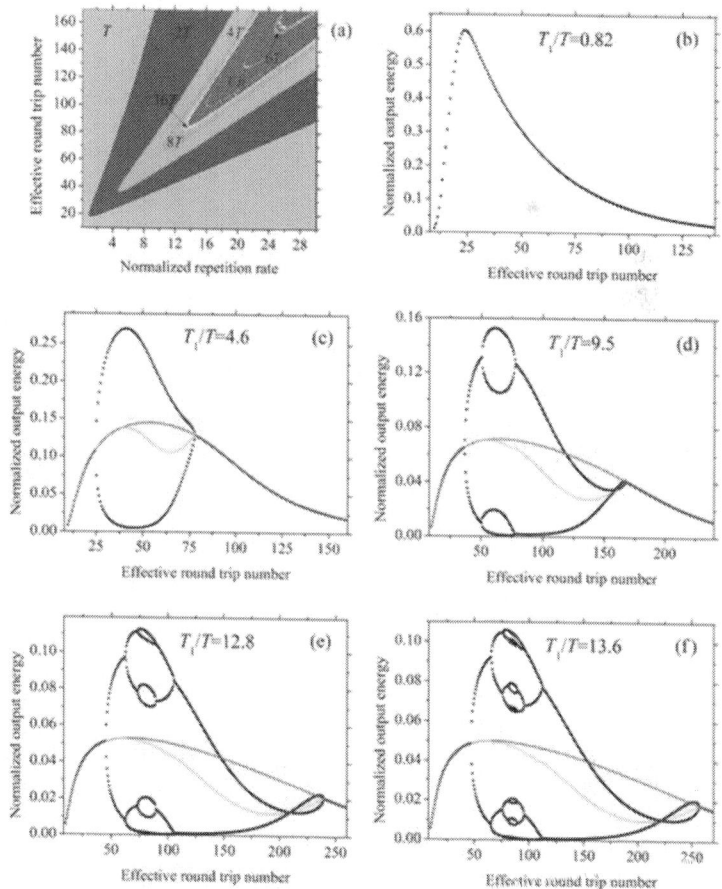

Figure 6. Diagram of dynamic regimes (a) and corresponding bifurcation diagrams for selected repetition rates (b–f) at $s=7.710^7$ and $g\,t=0.028$. Pulse energy, averaged pulse energy and "fixed point" energy correspond to black, green and red lines respectively.

We also determined real (accounting multi-energy nature) output energy averaged over large number of operation cycles $f(k)$ (Fig. 6). Interestingly, the real averaged energy is considerably lower than the reference "fixed point" energy in regimes exhibiting pronounced period doubling. The question is "what can the origin of this energy defect be?" One can suppose that in a multi-energy regime a relatively larger part of the resonator energy circulating is redistributed to the channel of parasitic losses. Some evidence for this explanation is that the same curves calculated in case of zero losses coincide in spite of bifurcations. The only alternate reason is stored energy depletion through spontaneous emission during the pump phase. However, the latter effect becomes dominant when the population inversion is on average high (i.e. at low round trips), that is not the case now.

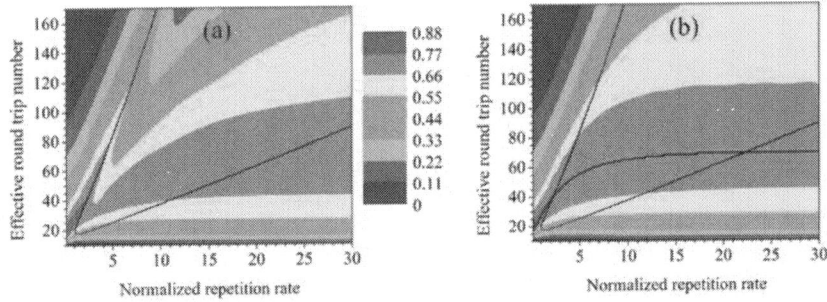

Figure 7. Efficiency of stored energy extraction with respect to the separatrix (a); the reference ("fixed point") efficiency with respect to the separatrix and *max* curve (b); both diagrams are for $s = 7.7 \cdot 10^7$ and $g\,t = 0.028$.

Single-valued functions characterizing output energy allow presentation of an informative picture of system performance in 2D diagrams. The extraction efficiency of stored population inversion may serve as a more convenient parameter for this purpose. The extraction efficiency can be defined as averaged output energy divided by maximum stored energy available at a given repetition rate, $f(k)/max$, where $max = 1\exp(T_1/T)$. Then the value calculated in the fixed point ($_1/max$) can be regarded as the maximum attainable efficiency at a given round trip number. Typical diagrams of these parameters in the space of repetition rate versus round trip number are presented in Fig. 7. Additional normalization to *max* introduced for already normalized terms is not meshing. Quite the contrary, it simplifies the practical use of theoretical data. The efficiency is the ratio which has the same value for real "dimensional" and normalized parameters.

Apparently, stable single-energy operation is the only suitable regime for routine use of regenerative amplifiers. For completeness sake we can note that one may successfully use certain unstable regimes for specific applications provided that there is comprehensive understanding of the essence of period doubling (Metzger et al., 2009). However this is rather the exception than the common rule. Thus, we can omit a detailed picture of dynamical regimes in order to move towards more pragmatic considerations. It is sufficient to leave only one curve in the parameter space defining the range of operating points in which operation is stable. This curve (further referred to as a separatrix) represents a manifold of the first order bifurcation points (1T-2T boundary) and actually separates zones of stable and unstable operation in the parameter space. Both real and "fixed point" energies are equal within a range of control parameters providing stable operation. Some drop of real efficiency with respect to that obtained in the assumption of stable operation is observed immediately below the upper branch of the separatrix, in the instability zone [Fig 7(a)]. The cross-sections of this feature are observed in Fig. 6 at several repetition rates and we have already concluded that this "valley" originates from enhanced influence of parasitic loses in the period doubling regime.

The distribution of the "fixed point" extraction efficiency ($_1/max$) in the parameter space contains sufficient information to determine system performance when accompanied with the separatrix curve. This curve confines the space of stable operation that is the range where the "fixed point" data correspond to reality. For system optimization it is important to find operation points potentially providing maximum performance at each given repetition rate. The manifold of such points represents a curve of round trip number versus repetition rate which is mapping the peak value of the "fixed point" extraction efficiency (2D distribution in parameter space) and further referred to as max curve [Fig 7(b)]. The potential performance becomes real at repetition rates in which the max curve is outside of the instability region and when the round trip number is set equal to *max*. If the *max* curve is inside the instability zone then the system capabilities are underexploited. The corresponding range of repetition rates can be called the "critical range". Within the critical range there are two possible positions of the operating point which may provide maximum output in the stable regime – the points along the lower and upper branch of the separatrix [see Fig 7(b)]. According to the diagram, at lower repetition rates in the critical range the upper branch has an advantage from an efficiency point of view. Operation at lesser round trip number (lower branch of the separatrix) becomes preferable at higher repetition rates. Thus optimization of the regenerative amplifier is actually reduced to selection of the round trip number which provides maximum output pulse energy and at the same time allows stable operation for the required repetition rate range (imposed by the system specifications). The corresponding round trip number is logical to call optimum (*opt*). Obviously, *opt* is equal to *max* outside the critical range.

Instructive inference of considered above regenerative amplification properties is that within critical repetition rates the round trip number takes on optimum value either near the lower or near the upper separatrix branch but always at the margin of unstable operation. In general, operation at the margin of stability incurs challenges for robust operation in real systems which undergo technical noises. Even slight changes to control parameters may result in system instability. Therefore reliably stable operation generally requires setting the operating parameters well away from the instability border, but this in turn leads to a reduction of the system performance.

There is an important parameter related to intracavity losses which may influence performance of regenerative amplifiers indirectly: the amount of intracavity energy dissipated during the amplification stage. Accumulated over round trips, the fraction of intracavity energy, dissipated through parasitic losses is subject to the specific operation regime. In particular, multiple passes of the already amplified optical pulse lead to substantial enhancement of energy dissipation. This, in turn, may give unacceptably high heating of intracavity components caused by the absorbed part of the dissipated power. One of the critical components in this respect can be the Pockels cell crystal. It may lose contrast under excessive heating possibly resulting in failure of regenerative amplifier operation. This effect is especially pronounced for systems intended for high power applications. The energy defect, arisen due to parasitic losses (l), can be determined as a product of intracavity energy, integrated over the amplification stage, and loss factor expressed by the threshold gain gt. Using this definition and Eqs. (4) and (7) we get:

$$\varepsilon l = gt \int_0^\tau \varepsilon(t\sim)dt\sim = \int_0^\tau \varepsilon(t\sim)g(t\sim)dt\sim - \varepsilon f = gai - gaf - \varepsilon f$$

$$(14)$$

where the integration variable $t\sim$ is the current normalized time. The same result can be obtained from the energy conservation law since during short (as we have initially assumed) amplification stage there are only two energy consumption channels – useful signal and parasitic losses. In case of multi-energy regime the effective lost part of the energy can be found by averaging:

$$\langle \varepsilon l \rangle = 1k \sum k(gai(k) - gaf(k) - \varepsilon f(k))$$

$$(15)$$

The diagram of dissipated energy (also normalized to maximum stored energy max), for the set of governing parameters used before, is presented in Fig. 8(a). We can see that the large number of passes, typical for the upper separatrix branch, substantially contributes to this parameter. So the dissipated energy is about 7.5 times higher at the repetition rate when theoretical efficiencies for both branches are equal (at $T_1/T6.5$). This increase with respect to the lower branch often makes operation at high round trip number, well above max point inefficient, despite theoretical preference.

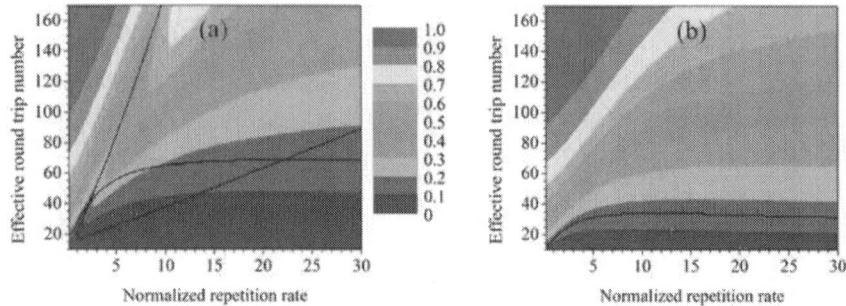

Figure 8. Dissipated intracavity energy, separatrix and *max* curves for $gt=0.028$ and $s=7.710^7$ (a); dissipated energy, and corresponding *max* curve for the seed energy $s=2.4610^4$ (b).

All the issues described in this section are regarded to one certain seed energy $s=7.710^7$. The corresponding diagrams give a typical but single section of multidimensional space of control parameters. However, as we can already conclude from amplification dynamics data presented in sections 2.6 and 2.7, the critical repetition rate range depends on the seed pulse energy. Thus, the picture of regenerative amplification is still incomplete and it is time to proceed to consideration of system optimization taking into account influence of the seed pulse energy.

Stability Diagrams and Pulse Duration Effects

It is possible to present data which allow evaluation of regenerative amplification of different seed pulse energies by considering a single diagram. The necessary premises for doing that have been formulated in the previous sections. The condition where bifurcations are absent in the whole parameter space (Table 1) gives a general understanding of the seed energy influence; however this condition is too strong from a practical point of view. In order to thoroughly utilize power capabilities of the regenerative amplifier at a certain repetition rate, the round trip number should be set equaled to*max*. Stable operation requires having the point *max* outside the instability zone (delimited by the separatrix curve on the parameter space). The inter-positioning of the separatrix and the *max* curve contains sufficient information to determine the critical repetition rate range for a certain seed energy and the possible optimum operation points (*opt*) within this range (which lie, as we have known, along one of the separatrix branches). The diagrams consisting of *max* curves and separatrixes (further referred to as stability diagrams) for selected pulse seed energies are presented in Fig. 9.

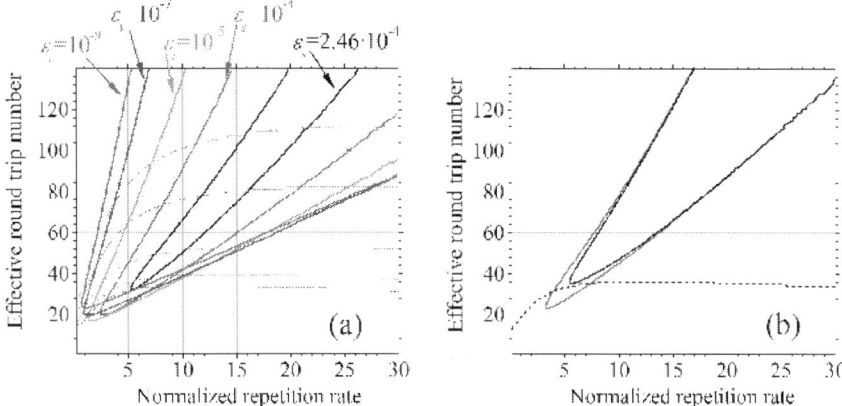

Figure 9. Stability diagrams for the threshold gain $gt=0.028$; the curves couples of the same color (the separatrix and *max* curve) correspond to certain seed energies (a). The separatrixes are shown for the short pulse (black solid line) and long pulse approximations (red solid line) and *max* curve (dash line) at $gt=0.028$ and $s=1.710^4$ (b).

The approach of stability diagrams forms a more systematic concept of the system behavior and specifically allows estimation of the seed level which may enable one to avoid instability effects at the required pulse repetition rate. Also we can see that the critical range shrinks as the seed energy is increasing. The most critical repetition rate is $T/T_1=5.5$, the point of the "worst stability", requiring the highest seed energy for optimal operation. Finally one can determine the seed pulse energy at which *max* curve does not pass instability zone at all. For the specific parasitic losses that we consider ($gt=0.028$) this energy is equal to 2.4610^4. Consequently, a regenerative amplifier seeded with pulse energy higher than that value (further referred to as "ample") allows theoretically attainable average power and stable operation over the whole repetition rate range. Also we can mention here another advantage of the "ample" seed operation: As soon as the critical range has disappeared there is no need to operate at the upper separatrix branch and to suffer from large intracavity energy dissipation peculiar to this regime. The diagram of dissipated energy for the ample pulse seed energy, $s = 2.4610^4$ is depicted in Fig. 8(b).

The approach of stability diagrams is a straightforward way of regenerative amplifiers optimization. However in real systems there are specific effects which influence dynamic behavior and output parameters and they cannot be elaborated by using only approximation of simplified rate equations. Nevertheless it is possible to understand some important contributions staying basically within present approach.

The major application of regenerative amplifiers is amplification of short (even more common term is ultrashort) optical pulses. Now we shall consider the influence of pulse duration on regenerative amplification. The theory described above is based on rate equations formulated for idealized four-level system. One of the positions in the definition of that "ideality" is instant depopulation of the lower laser level (which is also called terminal level). In reality we can assume that the lower laser level is virtually unpopulated, only provided that the amplified optical pulse is much longer than the terminal-level lifetime (long pulse approximation). Otherwise, if the length of the optical pulse is much less than the terminal lifetime then the terminal level will remain populated resulting in "faster" decay of population inversion (indeed during amplification of a single pulse). Comprehensive evaluation of actual terminal-level lifetime provided by (Bibeau, et al., 1995) for different neodymium doped laser media gives actual values well exceeding 100 ps. So we can conclude that amplification of pulses shorter than 100ps virtually for all neodymium based systems is more appropriate to analyze within the short pulse approximation that is assuming negligible terminal-level depopulation during singe pulse amplification. This, in fact, constrains applicability of presently described approach and leaves beyond the scope functionally important ultrashort pulses. We can note in advance that net contribution of the terminal-lifetime effect to regenerative amplifier behavior is rather weak at high repetition rates. However it gives some quantitative refinements to the picture presented above.

In order to elaborate regenerative amplification in the short pulse approximation we need to re-calculate the fundamental relations of final and initial gains gaf=gˆa(gai). In this approximation the amplification of a single pulse was regarded as that in truly three-level gain media with initially empty ground state. After single pass amplification the ground state becomes partially populated, but by the second pass it is empty again and consequently the gain defect, which appeared due to "instant" three-levelness, is recovered. The latter is the case since we assume the terminal-level lifetime to be much shorter than the round trip time, T_0. Typical T_0 value is in the range between ten and a few tens of nanoseconds, so this is good approximation for most of neodymium media [actually except some fluoride crystals and glasses in which the neodymium terminal lifetime is of 10 ns order (Bibeau, et al., 1995)]. By this means and by using sequential procedure the basic relations gaf=gˆa(gai) and corresponding output energies were determined. The subsequent procedures (evaluation of the fixed points and their stability analysis) stay unchanged for the short pulse approximation.

We revealed that the influence of a terminal level appears as follows. There is noticeable deviation between the *max* curves at low repetition rates. However both "short pulse" and "long pulse" curves virtually coincide at the repetition rates $T_1/T>1.0$, indeed in the range that we are studying. The separatrix curves practically coincide at low seed energy levels, although filling of the instability zone differs from what have been seen earlier [e.g. in Fig 6(a)] towards not so wide variety of regimes. These peculiarities add little in practical essence. Therefore we have come to nothing more than qualitative description and statement of that fact.

The only noticeable influence of the terminal lifetime was found at high seed energy, approaching to the ample level. The stability diagram, depicted in Fig.9 (b), shows what the difference is. In comparing two separatrixes we can point out that the tip of the "short pulse" separatrix is somewhat shifted towards higher round trip number and higher repetition rate. Such a shrink of the instability zone gives certain improvement of general system stability. The ample seed pulse energy determined for short pulses is almost 1.5 times less than that found within long pulse approximation ($s=1.710^4$ against 2.4610^4). This improvement is rather unexpected result since terminal-level "bottleneck" in some respect hampers the ideal four-level amplification. Thus at least partially and shortly populated terminal level does not act as additional losses as one might imagine.

Another phenomenon, substantial at high intensities and therefore requiring intent attention while ultrashort optical pulses are amplifying, is the Kerr effect. This nonlinearity makes an impact on amplification process by intensity dependent refractive index change in volume intracavity components. The influence occurs in both spatial and spectral domains and commonly is described as Kerr lensing and self-phase modulation. Reduction of this negative influence usually implies decrease of effective optical pulse intensity which can be quantified in terms of so-called B integral, in essence representing nonlinear on-axis phase shift (Brawn, 1981). In case of regenerative amplification multiple passes should be taken into account in order to evaluate total B integral accumulated during amplification stage:

$$B \approx B_1 \int_0^\tau \varepsilon(t\sim)dt\sim \qquad (16)$$

where B_1 is single pass B integral calculated for the intensity which is equal to the gain media saturation fluence divided by the pulse duration (see the Appendix section for details). The energy integral in this formula exhibits a factor of effective impact of multiple passes. We have already met such an integral when evaluated total lost part of the energy (see Eq. 14) and the value proportional to this factor has been depicted in previous section in Fig. 8. We can see that operation at the upper branch of the separatrix with short pulses is strongly unadvisable since the B integral is increased by several times. The cause of that is obviously multiple passes of intense pulses peculiar in a regime behind the instability zone.

That concerns the B_1 value, its reduction, in a condition of fixed pulse duration, simply implies standard methods of the mode area increase and shortening of the volume intracavity components. However these possibilities are rather limited. Even so the thin disc geometry allows tremendous reduction of gain media length (Speiser & Giesen, 2008) but the Pockels cell still can exhibit a real challenge. Among known to date Pockels cell materials only the BBO crystal is suitable for high average power (due to low absorption) and for high repetition rates (thanks to relatively low acoustic ringing). However the transverse electro-optic effect, the only functional for this crystal, permits shortening the optical pass or aperture increasing only in limited extent until driving voltage becomes unacceptably high (Nickel et al., 2005). Thus, since the Kerr effect often restricts the capabilities of regenerative amplifiers it is important at least to select correctly the operation regime in order to minimize its impact. We can note that operation at higher seed energy (*s* larger than ample seed pulse energy) is beneficial in this respect too. At that condition, operation at maximum output (*max* curve) does not suffer from too high multiple pass factor of the B integral [see Fig. 8(b)].

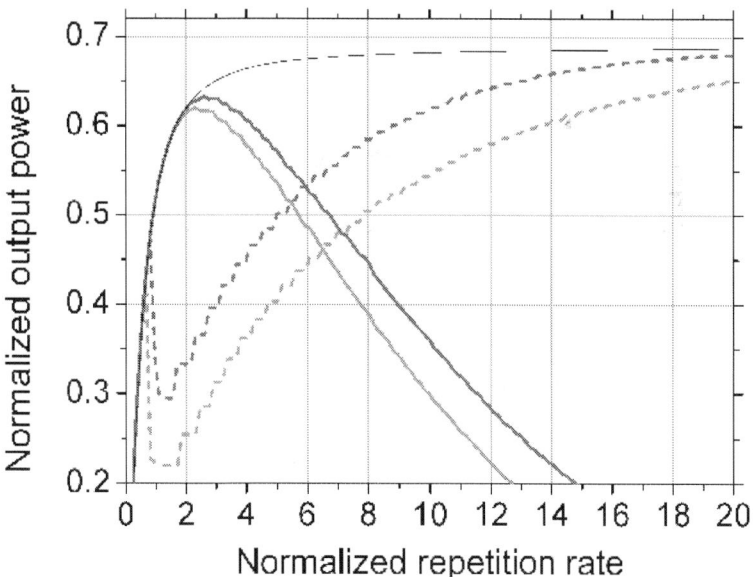

Figure 10. Theoretical output power versus repetition rate for the threshold gain of 0.028 and normalized seed pulse energy of $s = 7.710^7$ (red line) and $s = 7.710^9$ (green line). The data of high and low separatrix branches correspond to solid and dashed lines respectively. The black line is the reference curve of achievable power.

At the end of the theoretical part of the chapter we are giving diagrams of the basic performance parameter – maximum output power which can be achieved in a condition of stability maintaining (=*opt*), Fig. 10 . Presented normalized average power is defined as a product of normalized energy and normalized repetition rate. The power curve, calculated for *max* round trip number, can be regarded as the reference curve corresponding to theoretical efficiency limit. Really, in accordance with the *max* definition, the corresponding output power is the highest obtainable average power in assumption of period doubling absence. The calculations show that this power is invariant under the seed energy, although the *max* itself is not. Achieving the calculated maximum average power implies the best possible utilization of the stored pump energy, i.e. it assures the highest power efficiency. Obviously the same top performance we can reach operating at the ample seed energy provided that the round trip number is set equal to *max*. This attribute is the main benefit of the high seed pulse energy at high repetition rates. The power defect with respect to the reference power curve observed for lower seed energies is essentially caused by instability effects. We present curves related to both branches of the separatrix within critical repetition rate range. The data related to the upper branch are not always reliable because as we have shown earlier this regime suffers from indirect effects related to multiple passes of intense optical pulse such as the Kerr effect and excessive heating of optical components.

In concluding this theoretical part, now we are ready to proceed to experimental verification of the ideas developed above.

AMPLIFICATION EXPERIMENTS

Experimental Setup

The amplification experiments were carried out in order to demonstrate basic features of amplification dynamics and to verify theoretical results presented above. The knowledge of potential capabilities and of general limitations makes it possible to provide the best regime selection and deliberate optimization of control parameters. In practicality, this means to maximize extraction of the given stored population inversion as a stable train of output pulses. We leave outside the scope of present consideration optimization of the pump characteristics and the geometry of the optical resonator allowing more power in TEM_{00} mode, as these do not relate directly to the amplification dynamics. The parasitic intracavity losses, although formally a governing parameter, are not an object of consideration; they should simply be reduced as much as technically possible. Since the repetition rate is usually imposed by the specifications it appears as a variable but a given parameter. There are two adjustable parameters which can be used for the system optimization – the number of roundtrips and the seed pulse energy.

We performed experiments with a system based on $Nd:YVO_4$ crystals, a gain medium with truly four-level nature (except terminal-level nuances). The schematic diagram of the experimental setup that we used for investigation of regenerative amplification is shown in Fig. 11 . In essence the system consists of the seed source and the regenerative amplifier itself.

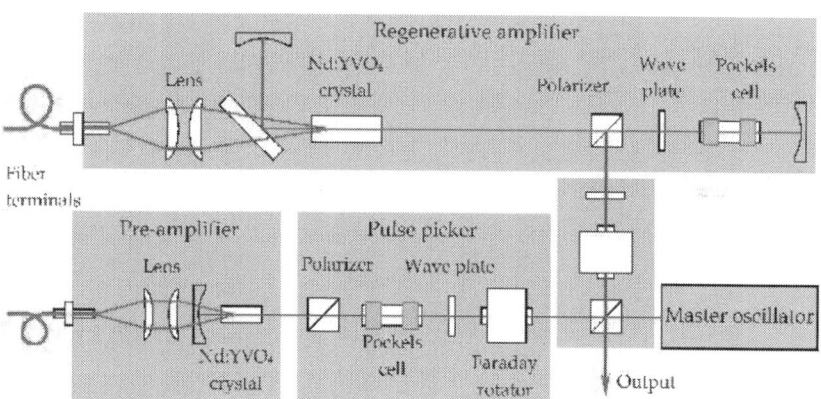

Figure 11. Schematic diagram of the experimental setup.

The seed pulse source for regenerative amplification experiments was based on a diode-pumped passively mode locked picosecond $Nd:YVO_4$ laser of moderate power. It generated a continuous pulse train with repetition frequency of 82 MHz and average power of 300 mW. The laser was able to produce optical pulses with duration as short as 6 ps. The short pulses were used in experiments where dynamics peculiar to high peak intensities were of interest. The initial investigations were focused on the "pure" dynamics not disturbed by optical nonlinearities. These experiments were carried out with 58 ps duration pulses obtained by installing an etalon in the oscillator cavity (the etalon narrows the bandwidth, thus widening the pulse duration).

A pulse picker was used to select pulses for further amplification and in this way to control the effective repetition frequency of the seed source. This part of the seed source system is important for high repetition rate operation, especially based on high gain laser media like neodymium doped vanadates. If the pulse picker is not used then two negative effects caused by unwanted pulses have place. These pulses continuously pass the optical resonator of the regenerative amplifier during pump stage and go out spatially coinciding with useful output signal. This leads first to reduced pulse contrast and second to parasitic consumption of stored energy. One can decrease the seed pulse energy and compensate for that by increasing number of round trips. This simple approach may often avoid bad influence of the unwanted seed background while operating at low repetition rates. The reduction of the seed energy, as we have already seen (theoretically), is not a good idea when turning operation to high repetition rates. In our setup the pulse picker was an electro-optic switch based on an RTP Pockels cell. Selected pulses formed an input signal for the preamplifier. The remaining pulses of the master oscillator train were directed to the fast photodiode for synchronization of electro optic components of the system including the pulse picker itself.

A double pass Nd:YVO$_4$ preamplifier installed behind the pulse picker was used to increase the seed pulse energy to required ample energy. High emission cross section of the Nd:YVO$_4$ crystal make this system efficient at relatively low input average power. Only 2 W of pumping was sufficient to achieve a gain coefficient of more than two orders of magnitude. The seed pulse energy was 3.2 nJ when pumping of the preamplifier was switched off. The energy of the pre-amplified pulse reached 1.1 μJ at 10 kHz and steadily decreased with the repetition rate to 370 nJ at 200 kHz. The calculation presented in the next section will show that the obtained energy is sufficient to ensure stable operation.

Simple estimations show that the preamplifier is a good alternative in comparison with a more straightforward seed source scheme based on a powerful master oscillator. In order to provide 370 nJ pulses the master oscillator operating in a CW mode locking regime with a reasonable repetition frequency of 50 MHz should generate average power of 18.5 W. This way is really prodigal since the useful part of this power is much lower, e.g. only 74 mW even operating at 200 kHz. Obviously the preamplifier is a much more energy-efficient solution.

The regenerative amplifier was comprised of an optical resonator containing the gain medium (Nd:YVO$_4$ crystals) and an electro-optic switch. The electro-optic switch consisted of a BBO Pockels cell, a quarter-wave plate and a thin-film polarizer. The total multi-pass gain of the regenerative amplifier depends on the number of cavity round trips which is determined by the amplification-stage duration. This important parameter is easily controlled by setting the time interval during which the high voltage is applied to the Pockels cell.

The laser crystal was continuously pumped by the fiber coupled laser diode module with fiber core diameter of 400 μm and numerical aperture of 0.22. Optimal pump power, providing maximum output in TEM$_{00}$ mode, was set to be 44 W. This optimization was performed in CW generation mode. Provided that no voltage was applied to the Pockels cell and the quarter-wave plate was adjusted for maximum output (optimal output coupling conditions), 12.5 W of average power was obtained.

The output radiation was diverted from the input signal path by standard optical circulator based on Faraday rotator. The repetition rate of the system was limited to 200 kHz by electronics driving the electro-optic switches.

Application of Stability Diagrams to Amplification Experiments

Now we can apply the concepts developed in the theoretical part to estimating behavior of a real system. At the beginning the basic system properties should be evaluated in order to perform reciprocal transformation of normalized parameters to dimensional ones corresponding to real operation conditions. The steady state small signal gain, determined by the pump intensity, was found directly as the ratio of the seed energies measured right before and behind the active element of the regenerative amplifier. The parasitic intracavity losses were derived from the specifications of the optical components. The laser characteristics of Nd:YVO₄ crystal (emission cross section and gain relaxation time) were taken from (Peterson et al., 2002). We explored amplification of three seed pulse energies differing by about two orders of magnitude: pre-amplified seed, unamplified seed and attenuated seed. This set of input signals covers the functionally important range. The value of the unamplified seed, 3.2 nJ, is of the same order of magnitude as the pulse energy of commonly used moderate power solid-state picosecond lasers. Operation with the seed energy intentionally attenuated to 32 pJ provides opportunity to evaluate typical behavior of the regenerative amplifier seeded with potentially attractive low power sources, e.g. with ultrafast laser diodes, which would substantially reduce system size and complexity. The seed energy obtained with the pre-amplifier was expected to be high enough to reach ample level. These seed energies were measured at the output of the seed formation system. However we observed that during further amplification the seed energy was not completely exploited. Mode mismatching reduces the effective seed energy. In general, it is difficult to avoid mode mismatching between a seed laser and the optical cavity of the regenerative amplifier in both spatial and spectral domains. Spectral mismatching can exist even with identical gain media because of e.g. different temperatures of laser crystals in those devices (Murray & Lowdermilk, 1980).

Table 2. Parameters used for stability diagrams calculation.

Parameter	Value
Wavelength	1064 nm
Emission cross section	11.4 10 19 cm 2
Gain relaxation time	83 μs
E ffective mode diameter in the laser rod	
Steady state small signal gain	2.94
Seed pulse energies	11 pJ (low); 1.1 nJ (medium); 240?nJ (high)

In order to have appropriate data for theory verification we experimentally estimated the overall level of mismatching. We measured real double pass small signal gain of the whole regenerative amplifier (output/input energy ratio while the Pockels cell was disabled) and compared this value with "pure" steady state gain of the laser element. Thus we determined the effective seed value to be about three times less than the measured one. The primary parameters of the regenerative amplifier which we used for calculations are summarized in Table 2.

Before proceeding to stability diagrams describing our particular experimental conditions, we note that we utilized a linear cavity that establishes double pass through the gain media, while the modeling has implied single pass for one cavity round trip. This factor of two was taken into account so that the y-axis of stability diagrams corresponds to real round trip number, $t/T_0 = \frac{1}{2}/G_0$. The other relations between dimensional and normalized parameters remain unchanged as they have been given in the section 2.2.

The stability diagrams describing amplification experiments are presented in Fig. 12. These data were obtained in the approximation of short pulses (accounting for terminal-level lifetime effect). As we have already known, the system capabilities are completely exploited when the *max* curve is outside the instability region. The diagrams show that at low seed energy (11 pJ) the appropriate repetition rates should be less than 20 kHz. For the medium seed level (1.1 nJ) this range increases to 25 kHz. At higher repetition rates the *max* curve enters the instability zone. It is important that for 240 nJ or higher seed energies the *max* curve does not enter the instability zone in the whole range of repetition rates. So, for our laser system this energy corresponds to the ample seed value, sufficient to eliminate negative features of amplification dynamics, and thus to completely exploit the system capabilities.

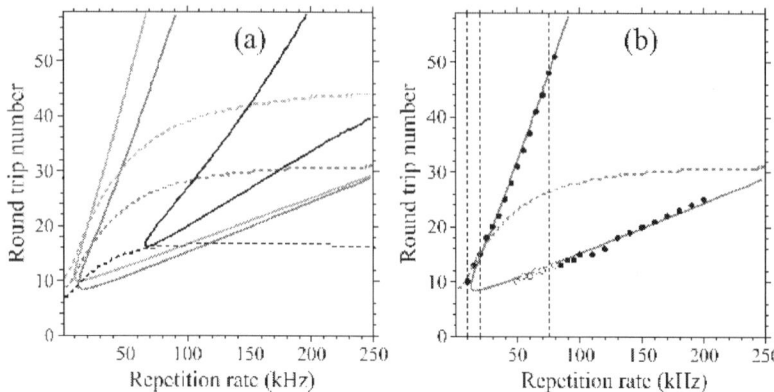

Figure 12. a) Separatrix curves (solid lines) and *max* curves (dotted lines) in parameter space. Black, red and green lines correspond to seed pulse energies of 240 nJ, 1.1 nJ and 11 pJ, respectively. (b) Operating point trajectories (vertical dashed lines) and measured number of optimal round trips for the pulse durations of 58 ps (solid circles) and 9 ps (open circles) with respect to stability diagram for the seed energy of 1.1 nJ.

Experimental Bifurcation Diagrams

The initial experiments were carried out with medium seed pulse energy (the preamplifier was disabled). Various dynamic regimes, depending on set of control parameters, were observed. As an illustration, Fig. 13 shows oscilloscope screen shots of the output pulse train in typical single-energy and period doubling regimes.

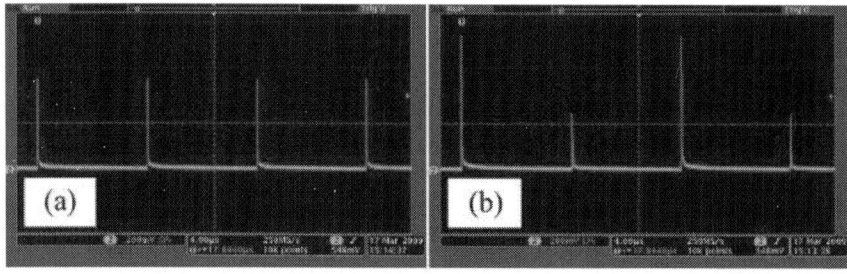

Figure 13. Screenshots of typical pulse trains at 90 kHz. Stable energy output and the 2T period doubling regime were obtained at number of round trips equal to 14 (a) and 16 (b) respectively.

Experimentally obtained diagrams of the average output power and pulse energy versus number of round trips demonstrating system behavior at different repetition rates are presented in Fig. 14. The specific repetition rates were chosen to describe the most relevant cases of regenerative amplifier dynamics in respect of system optimization. The single-peaked dependence inherent to low repetition rates appears at 10 kHz [Fig. 14(a)]. The average power and the pulse energy reach the maximum values simultaneously, when the round trip number is equal to ten. At 20 kHz the situation is different [Fig. 14(b)]. The shape of the energy curve shows that the system undergoes bifurcation in the 9–13 range of the round trip numbers. However, in this case the period doubling does not affect the system performance because the output power reaches its maximum value in a single-energy regime. This repetition rate is still not critical.

Instability effects become more pronounced at higher pulse repetition rates. The period doubling not only breaks the energy stability but also distorts the curve of the average power (as described in section 2.8). This curve now has two explicit peaks [Figs. 14(c) and 15(d)]. The first peak, corresponding to the maximum power, is located in a period doubling zone, whereas the second one is just over the instability edge. The optimal regime is obtained in the vicinity of the bifurcation point. At 75 kHz the optimal number round trips is equal to 48. This point is close to the second power peak, on the right side of the period doubling zone [Fig. 14(c)]. For 90 kHz repetition rate the optimal number round trips is equal to 13 and is situated right before the first bifurcation point [Fig. 14(d)].

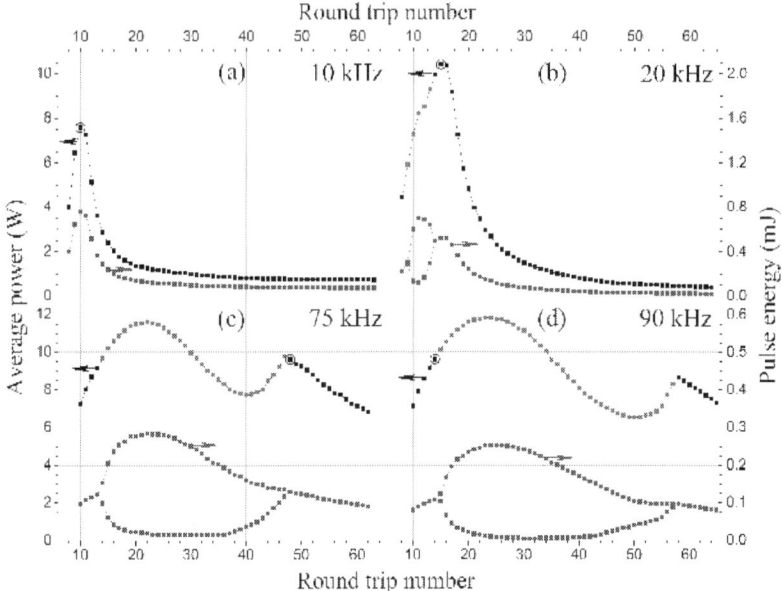

Figure 14. Experimental average power (black and red dots correspond to stable and unstable regimes respectively) and pulse energy (blue dots) versus number of cavity round trips for the selected repetition rates. The encircled points correspond to the maximum power at stable operation.

The trajectories of the operating points corresponding to variation round trip number at a constant repetition rate are presented in Fig. 12(b).This trajectory at 10 kHz does not pass the instability zone. At 20 kHz the optimal operating point is above the instability zone. Both repetition rates 75 kHz and 90 kHz are critical – the optimal number of round trips is on the stability edge and is rather far from the point of the highest attainable power. The experimentally observed results confirm theoretical predictions that for critical repetition rates: (i) the output energy exhibits unacceptable fluctuations when the amplifier produces the highest average power, (ii) the highest stable pulse energy is reached close to the instability edge.

Some effects caused by nonzero terminal-level lifetime were also observed in the experiment. The presented bifurcation diagrams at 75 kHz and 90 kHz theoretically should exhibit not only 2T but also 4T regimes [like in Fig. 6(e)], if the long pulse approximation is applied. The regime of 4T period doubling was not observed experimentally and it should not exist theoretically, provided that contribution of the terminal-level lifetime is accounted. At the same time, as the theory has predicted, terminal-level lifetime effect does not influence the system performance at such low seed level.

Real deviation from theory is observed at the lowest repetition rate (10 kHz). The output energy decays too fast behind the peak point in comparison with theoretical expectations [see Fig. 6(b)]. This occurred because of the Kerr effect influence was substantial at low repetition rates even for initial experiments performed with relatively long 58 ps pulse duration.

Performance Evaluation and Discussion

It has been shown in the theoretical part that variation of only roundtrip number does not solve the stability problem; increase in the seed pulse energy is required in order to avoid bifurcations and corresponding instability at high repetition rates. So, we proceed to experimental verification of the seed energy influence. We compared operation for three cases: "low", "medium", and "ample" pulse seed energy.

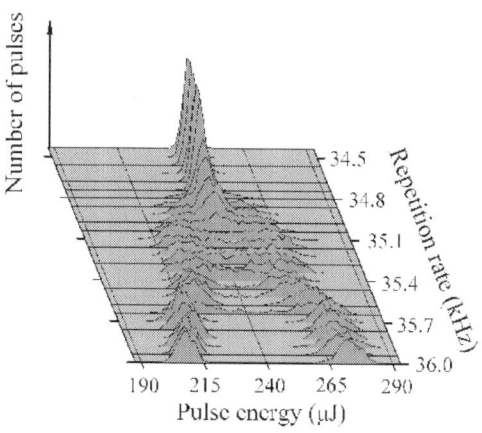

Figure 15. Experimental energy histogram in the vicinity of the bifurcation point.

The measurements were performed at optimal round trip numbers. They were set for maximum average power while maintaining stable operation for every repetition rate. The formulation of stability criteria in real experimental environments requires certain attention. Experimental discrimination of amplification regimes to some extent suffers from uncertainty because of technical instabilities, not related to fundamental system properties. Technical noises, in essence slight modulation of governing parameters usually with random distribution, limit system stability. The most typical of these are pumping source noises, seed pulse energy fluctuation, synchronization jitter, resonator disturbance by mechanical vibrations and by air flow. In our setup the technical, not disturbed by period doubling, standard deviation of output energy was less than 0.7%. In proximity to a bifurcation point the deviation increased. The diagram of typical transition from single energy to period doubling regime "under magnification" of the repetition rate scale is shown in Fig. 15; the bifurcation point looks rather as a spot than as a point. The uncertainty takes place in the range where the deviation of energy is clearly higher than the technically conditional level but where the two peaks are still not distinguishable. In the repetition rate scale this range does not exceed 1 kHz (typically the value is 0.5 kHz). In the round trips scale measurements with such fine steps are not possible (in contrast to theoretically grounded terms the real experimental round trips are discrete), and besides, the analysis of the energy histogram is a time consuming procedure. So, we used simple phenomenological criterion formulated for the particular setup allowing real time measurements. The operation was considered stable when the standard deviation of the pulse energy did not exceed 1%. Apart from that, special care was taken regarding origin and spectrum of disturbing noises to be sure that they are virtually non-periodic. In particular the pulse picker driving electronics had to be improved in order to eliminate seed train modulation at the frequency equal to half the system repetition rate. We can note that even barely perceptible presence of "resonant" components (sub-harmonics of dumping frequency) in the spectrum of technical noises may tremendously enhance the influence of period doubling and can change the dynamical pattern beyond recognition.

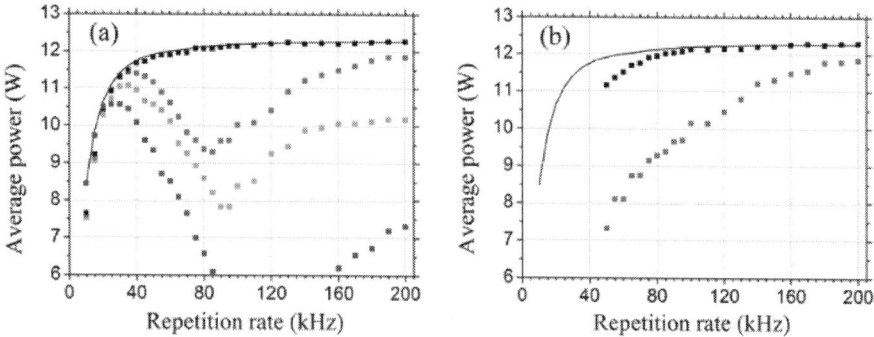

Figure 16. Experimental output power versus repetition rate for 58 ps (a) and 9 ps (b) pulses. Black, red and green dots correspond to measured seed pulse energies of 700 nJ, 3.2 nJ and 32 pJ, respectively. Theoretical curve of achievable power is solid line in both diagrams. Blue dots in Fig. 16 (a) are Q-switch experiment results.

Experimental dependences of the average output power versus repetition rate are presented in Fig. 16(a).The output is virtually independent of the seed level for low repetition rates. At higher rates there is a drop in power for low and medium seed levels in compliance with theoretical notions. The most significant power decrease appears in the 80–95 kHz range, then the output power steadily grows as the repetition rate increases.

Non-monotonic behavior of the power curve originates from the specific location of optimal operation points in the parameter space. Corresponding experimental data with respect to the theoretical stability diagram for medium seed energy is presented in Fig. 12(b).The operating points coincide with the theoretical curve of *max* until the latter enters into the instability zone (at 25 kHz). Then operation at the upper branch of the separatrix becomes optimal. However as the repetition rate increases the stable operating point moves further away from the *max* position, resulting in lower output energy. Consequently, starting from 85 kHz the optimal operation point switches to the lower border of the instability region. In this regime the operating point gradually comes closer to the *max* position, and consequently, the output power steadily increases with increasing repetition rate.

The pre-amplified seed pulse has sufficient energy to maintain stable operation at maximum power, in accordance with ample seed properties. The power curve obtained with the preamplifier shows no signs of downward excursion with respect to the reference curve (the curve of theoretically attainable power corresponding to *max* round trip number). Some slight deviation observed at low repetition rates was assumed to be caused by the Kerr effect. This nonlinearity can cause additional intensity-dependent intracavity losses (more pronounced at higher pulse energies), and so its influence is more pronounced at lower repetition rates. In order to differentiate the Kerr effect influence we performed experiments with nanosecond pulses in the same setup. The seed source was disabled so that the regenerative amplifier was transformed to a Q-switched laser with cavity dumping. Basically this device can be regarded as a regenerative amplifier seeded by spontaneous emission getting into the lasing mode. Thus, the seed energy becomes extremely low and additionally the laser system turns to the nanosecond domain (15 ns output pulse length was obtained in our setup). The former gives a large drop in power at critical repetition rates, the latter gives enhanced, with respect to the regular regenerative amplification, output energy at low repetition rate (the Kerr effect is negligible for such long pulses). The nanosecond pulse energy at low repetition rates exceeds the picosecond energy and very well agrees to the theoretical curve [Fig. 16(a)]. This demonstrates that experimental deviations appeared because of the Kerr effect and consequently general validates that the theoretical approach we have developed is wholly satisfactory.

Heating of intracavity components was not observed to affect regenerative amplifier operation in optimal round trip number experiments. However, clear evidence of excessive heating was noticed during recording of bifurcation diagrams in the worse (from this point of view) regime. The beam quality was observed to deteriorate for 90 kHz repetition rate and at round trip numbers between 46 and 56. From the other hand the average power defect reached maximum in this range [see Fig. 14(d)]. As we have concluded theoretically, this defect is indeed due to dissipated power, the absorbed fraction of which is heating intracavity optics.

We also compared performances of the regenerative amplifier seeded by high and medium pulse energy for the functionally important case of shorter optical pulses. Typically about 9 ps pulse duration was obtained at the output of the amplifier seeded by the 6 ps pulse. This duration is close to the minimum value supported by the gain bandwidth of the Nd:YVO$_4$ crystal in high total-gain applications such as regenerative amplification. The measurements were constrained to dumping rates above 50 kHz. Nevertheless, the intensities were substantial, and the Kerr effect influence was so strong that it eventually resulted in decrease of the output power [Fig. 16 (b)]. The average power obtained with the pre-amplified seed was slightly lower than that theoretically predicted below 80 kHz and the difference reached 6.7% at 50 kHz. However, the comparison of these characteristics with those obtained at a medium seed level shows that the benefit of the preamplifier is even more pronounced in case of shorter pulses. The difference is related to a large decrease of the average power below 85 kHz for the case of the medium seed. This is direct consequence of inefficient operation at the upper separatrix branch in critical repetition rate range under the Kerr effect influence. Experimental optimal operation points for both 58 ps and 9 ps pulses with respect to theoretical stability diagram are represented in Fig. 12(b). The optimal operating points for the short pulse experiments were always settled along the less efficient lower branch of the parameter separatrix. Attempts to operate at the upper branch (optimum for long pulses) resulted in an even larger decrease of the output. In order to quantify this difference we estimated the multi-pass B integral of the system (Eq. 22 in the Appendix). The B integral calculated at 50 kHz repetition rate gave values of 1.3 (acceptable) and 7.6 (problematic) at the transition from low to upper separatrix branch respectively. Thus the high seed energy gives additional advantage at shorter pulses due to a significantly lower value of the optimal number of round trips.

The amplification experiments which were performed with Nd:YVO$_4$ regenerative amplifier have shown that the developed theoretical approach accurately agrees with experimental data and can be used for practical system-design guidelines.

CONCLUSIONS

Continuously pumped regenerative amplifiers are subject to energy instability at high pulse repetition rates due to period doubling bifurcation. Theoretical concepts representing a generalized picture of operation features have been in-detail worked out in order to differentiate and understand instability effects. Experimental data for Nd:YVO$_4$ regenerative amplifier have been presented; and possible techniques for performance optimization have been analyzed. An increase in the seed pulse energy has been demonstrated to improve amplification dynamics. Addition of a preamplifier is shown to be not only a convenient means for regenerative amplification investigation at a wide range of seed energies but also an efficient way to get top performance in practice. The Nd:YVO$_4$ preamplifier delivered seed energy high enough to provide stable operation at repetition rates up to 200 kHz with average output power near the theoretical limit.

We have not performed appropriate experiments to elaborate amplification dynamics for ytterbium based regenerative amplifiers. However, it would be interesting to verify theoretical predictions for ytterbium doped crystals as for media exhibiting pronounced quasi-three-level behavior. Especially, because similar types of crystals may exhibit different sensitivity to bifurcations in the same setup, as reported by (Buenting et al., 2009) and (Sayinc et al., 2009). In addition, bulk ytterbium doped materials are low gain materials and therefore the preamplifier technique is not so easily applicable to such systems. There is demand to create another method to keep regenerative amplifiers stable at maximum output power, possibly some kind of feedback. Such an idea was formulated by (Dörring et al., 2004) but as far as we know neither theoretical modeling nor practical realization of this approach has been reported so far. Essentially, this can be an attempt to stabilize the inherently unstable balance between pumping and inversion depletion – that is constant forced return of the system state to the originally repulsive fixed point. Actually the initial gain of amplification phase should be kept constant by the feedback. However, straightforward engineering does not work in this case because output energy signal of the current operation cycle does not contain sufficient information to adequately control pump power of the subsequent cycle. So elaboration of this problem can be a goal for further theoretical research and experimental work.

Appendix. Multi-pass B Integral for Regenerative Amplifier

The conventional quantitative gauge of the Kerr effect in laser systems is the B integral, nonlinear on-axis phase shift which light waves with wavelength undergo propagating through the media:

$$Bsp=2\pi\lambda\int n2(z)I(z)dz \qquad (16)$$

The terms $I(z)$ and $n_2(z)$ are distributions of on-axis intensity and nonlinear refractive index along current coordinate z. In order to evaluate the Kerr effect accumulated during regenerative amplification the integration should be performed over all the roundtrips of the optical cavity (multi-pass B integral). Then the full integration length is a product of the optical cavity pass length and round trip number. In an approximation of relatively low single pass gain (the integral within the gain medium can be replaced with the average) and also for moderate Kerr effect influence (iteration of the intensity profile in the optical resonator is not disturbed too much by self focusing) we can replace the overall integral with a sum of single pass integrals:

$$B\approx2\pi\lambda\sum NRT\int n2(z)INRT(z)dz \qquad (18)$$

where index NRT implies summation over round trips. The sum of integrals is equal to the integral of the sum and also $n_2(z)$ is independent of round trip number function, then we obtain:

$$B\approx2\pi\lambda\int n2(z)\sum NRTINRT(z)dz \qquad (19)$$

In an assumption of Gaussian beam shape for which the peak intensity is equal to $2P/(w^2)$ (the term P is the optical power, the term w is the Gaussian beam radius) we can calculate the intensity for the pulse duration t in terms of pulse energy:

$$\sum NRTINRT(z) \approx 2\pi w2(z)\Delta t \sum NRTENRT$$

(20)

Summation of energies can be rewritten as time integration, and then dimensional current energy can be represented in terms of normalized energy which we have introduced in section 2.2:

$$\sum NRTENRT \approx 1T0 \int 0NRT \cdot T0E(t)dt = AaFsatG0T0 \int 0NRT \cdot T0\varepsilon(t)dt = AaFsat \int 0\tau\varepsilon(t\sim)dt\sim$$

(21)

At that, the limit of integration $NRTT_0$ (overall pulse propagation time) is reducing to the effective round trip number when the integration variable, current dimensional time t, has been transformed to the normalized time $t^\sim$. Also, the laser material properties, responsible for stimulated emission, are combined to a conventional macroscopic term, saturation fluence, the ratio of photon energy and stimulated emission cross section: $Fsat = h/$. The beam area in the active medium obviously can be expressed through the Gaussian beam radius: $Aa = w^2 a$. And finally, using straightforward transformation and normalized energy integral calculated in section 2.8 (Eq. 14), we derive explicit expression of B integral, the diminished form of which has been used in section 2.9 (Eq. 16):

$$B = B1 \int 0\tau\varepsilon(t\sim)dt\sim$$

where

$$B1 = 4\pi\lambda Fsat\Delta t \int n2(z)[waw(z)]2dz$$

And

$$\int 0\tau\varepsilon(t\sim)dt\sim = gaf - gai - \varepsilon fgt$$

(22)

We have obtained the multi-pass B integral for evaluation of the Kerr effect in regenerative amplifiers as the product of two factors. The first factor, B_1 represents attributes of the system geometry, material parameters and optical pulse duration. Essentially, this term is single-pass

B integral calculated for the Gaussian beam in given optical cavity and for pulse energy fluence equal to the gain medium saturation fluence. The second factor is a function of regenerative amplifier regime represented in normalized terms (the dissipated energy divided by the threshold gain). This is convenient for practical application form in which functional physical contributions are separated.

ACKNOWLEDGEMENTS

The authors wish to acknowledge the technical assistance of Juozas Verseckas from EKSPLA UAB in preparation of the experimental setup, Vidmantas Gulbinas from Institute of Physics and Lucian Hand from Altos Photonics Inc. for fruitful discussions of the manuscript. This work was partially financed by the Eurostars Project E!4335-UPLIT.

REFERENCES

1. K. Alligood, T. Sauer, J. Yorke, 1996 Chaos. An Introduction to Dynamical Systems, Springer-Verlag, 0-37894-677-2York
2. F. Arecchi, R. Meucci, G. Puccioni, J. Tredicce, 1982 Experimental Evidence of Subharmonic Bifurcations, Multistability, and Turbulence in a Q-Switched Gas Laser. Phys. Rev. Lett., 49 17 October 1982), 1217 1220 , 0031-9007
3. C. Bibeau, S. Payne, H. Powell, 1995 Direct measurements of the terminal laser level lifetime in neodymium-doped crystals and glasses. J. Opt. Soc. Am. B, 12 10 October 1995), 1981 1992 , 0740-3224
4. S. Biswal, J. Itatani, J. Nees, G. Mourou, 1998 Efficient energy extraction below the saturation fluence in a low-gain low-loss regenerative chirped-pulse amplifier. IEEE J. Sel. Top. Quantum Electron., 4 2 March 1998), 421 425 , 1077-260X
5. D. Brawn, 1981 High Peak Power Nd:Glass Laser Systems, Springer-Verlag, 0-38710-516-6York
6. U. Buenting, H. Sayinc, D. Wandt, U. Morgner, D. Kracht, 2009 Regenerative thin disk amplifier with combined gain spectra producing 500 μJ sub 200 fs pulses. Opt. Express, 17 10 May 2009), 8046-8050, 1094-4087

7. D. Clubley, A. Bell, G. Friel, 2008 High average power Nd:YVO based pico-second regenerative amplifier. Proc. SPIE, 6871 February 2008), 68711D, 0277-786X

8. J. Dörring, A. Killi, U. Morgner, A. Lang, M. Lederer, D. Kopf, 2004 Period doubling and deterministic chaos in continuously pumped regenerative amplifiers. Opt. Express, 12 8 April 2004), 1759 1768 , 1094-4087

9. M. Fermann, A. Galvanauskas, G. Sucha, 2002 Ultrafast Lasers: Technology and Applications, Marcel Dekker, 0-20391-020-6 York

10. S. Forget, F. Balembois, P. Georges, P. Devilder, 2002 A new 3D multipass amplifier based on Nd:YAG or Nd:YVO4 crystals. Appl. Phys. B, 75 4-5 (October 2002), 481-485, 0946-2171

11. M. Grishin, V. Gulbinas, A. Michailovas, 2007 Dynamics of high repetition rate regenerative amplifiers. Opt. Express, 15 15 July 2007), 9434 9443 4-9443, 1094-4087

12. M. Grishin, V. Gulbinas, A. Michailovas, J. Verseckas, 2008 Operation Features of Regenerative Amplifiers at High Repetition Rate, Technical digest in CD, paper CFB7, Conference on Lasers and Electro-Optics, CLEO-08, San-Chose, CA, USA, May 4-9, 2008, OSA

13. M. Grishin, V. Gulbinas, A. Michailovas, 2009 Bifurcation suppression for stability improvement in Nd:YVO4 regenerative amplifier. Opt. Express, 17 18 August 2009), 15700-15708, 1094-4087

14. H. Haken, 1975 Analogy between higher instabilities in fluids and lasers. Physics Letters A, 53 1 (May 1975) 77 78 , 0375-9601

15. Y. Jeong, J. Sahu, D. Payne, J. Nilsson, 2004 Ytterbium-doped large-core fiber laser with 1.36 kW continuous-wave output power. Opt. Express, 12 25 December 2004), 6088 6092, 1094-4087

16. J. Kawanaka, K. Yamakawa, H. Nishioka, K. Ueda, 2003 30-mJ, diode-pumped, chirped-pulse Yb:YLF regenerative amplifier. Opt. Lett., 28 21 November 2003), 2121 213 , 0146-9592

17. J. Kleinbauer, R. Knappe, R. Wallenstein, 2005 13-W picoseconds Nd:GdVO4 regenerative amplifier with 200-kHz repetition rate. Appl. Phys. B, 81 2-3 (July 2005), 163-166, 0946-2171

18. J. Kleinbauer, D. Eckert, S. Weiler, D. Sutter, 2008 80 W ultrafast CPA-free disk laser, Proc. SPIE, 6871 February 2008), 68711B, 0277-786X

19. W. Koechner, 2006 Solid-State Laser Engineering, Springer, 978-0-38729-094-2 USA

20. C. Liu, T. Riesbeck, X. Wang, J. Ge, Z. Xiang, J. Chen, H. Eichler, 2008 Influence of spherical aberrations on the performance of dynamically stable resonators, Optics Communications, 281 20 October, 2008) 5222 5228 0030-4018.

21. H. Liu, C. Gao, J. Tao, W. Zhao, Y. Wang, 2008 Compact tunable high power picosecond source based on Yb-doped fiber amplification of gain switch laser diode. Opt. Express, 16 11 May 2008), 7888-7893, 1094-4087

22. E. Lorenz, 1963 Deterministic nonperiodic flow. Journal of the Atmospheric Sciences, 20 2 March 1963), 130 141 , 0022-4928

23. W. Lowdermilk, J. Murray, 1980 The multipass amplifier: theory and numerical analysis. J. Appl. Phys., 51 5 May 1980), 2436 2444 , 0021-8979

24. V. Magni, 1986 Resonators for solid-state lasers with large-volume fundamental mode and high alignment stability. Applied Optics, 25 1 January 1986), 107 118 0003-6935

25. I. Matsushima, H. Yashiro, T. Tomie, 2006 10 kHz 40 W Ti:sapphire regenerative ring amplifier. Opt. Lett., 31 13 July 2006), 2066 2068, 0146-9592

26. J. Meijer, K. Dub, A. Gillner, D. Hoffmann, V. Kovalenko, T. Masuzawa, A. Ostendorf, R. Poprawe, W. Schulz, 2002 Laser Machining by short and ultrashort pulses, state of the art and new opportunities in the age of the photons. CIRP Annals- Manufacturing Technology, 51 2 February 2002), 531 550 , 0007-8506

27. T. Metzger, A. Schwarz, C. Teisset, D. Sutter, A. Killi, R. Kienberger, F. Krausz, 2009 High-repetition-rate picosecond pump laser based on a Yb:YAG disk amplifier for optical parametric amplification. Opt. Lett., 34 14 July 2009), 2123-2125, 0146-9592

28. G. Mourou, D. Umstadter, 1992 Development and Applications of Compact High-Intensity Lasers. Phys. Fluids B, 4 7 July 1992), 2317-2325, 0899-8213

29. D. Müller, A. Giesen, H. Hügel, 2003 Picosecond thin-disk regenerative amplifier. Proceedings of SPIE, 5120 November 2003), 281 286 , 0277-786X

30. J. Murray, W. Lowdermilk, 1980 Nd:YAG regenerative amplifier. J. Appl. Phys., 51 7 July 1980), 3548-3555, 0021-8979

31. D. Nickel, C. Stolzenburg, A. Bevertt, A. Geisen, J. Haüssermann, F. Butze, M. Leitner, 2005 200 kHz electro-optic switch for ultrafast laser systems. Rev. Sci. Instrum., 76 3 March 2005), 033111 033117 0034-6748

32. T. Norris, 1992 Femtosecond pulse amplification at 250 kHz with a Ti:sapphire regenerative amplifier and application to continuum generation. Opt. Lett., 17 14 July 1992), 1009 1011 , 0146-9592

33. R. Peterson, H. Jenssen, A. Cassanho, 2002 Investigation of the spectroscopic properties of Nd:YVO4. In: Proc. OSA TOPS, Advanced Solid-State Lasers, M.E. Fermann and L.R. Marshall, (Ed.), 68 294-298, OSA

34. A. Pugžlys, G. Andriukaitis, A. Baltuška, L. Su, J. Xu, H. Li, R. Li, W. Lai, P. Phua, A. Marcinkevičius, M. Fermann, L. Giniūnas, R. Danielius, S.Ališauskas, 2009 Multi-mJ, 200-fs, CW-pumped, cryogenically cooled, Yb,Na:CaF2 amplifier. Opt. Lett., 34 13 July 2009), 2075-2077, 0146-9592

35. G. Raciukaitis, M. Grishin, R. Danielius, J. Pocius, L. Giniūnas, 2006 High repetition rate ps- and fs- DPSS lasers for micromachining. Congress Proceedings (in CD), 99 Paper M1001, 0-91203-585-4 Congress on Applications of Lasers & Electro- Optics, ICALEO 2006, Scottsdale, USA, October 30- November 2, 2006, Laser Institute of America

36. F. Röser, T. Eidam, J. Rothhardt, O. Schmidt, D. Schimpf, J. Limpert, A. Tünnermann, 2007 Millijoule pulse energy high repetition rate femtosecond fiber chirped-pulse amplification system. Opt. Lett., 32 12 December 2007), 3495 3497 , 0146-9592

37. I. Ross, G. New, P. Bates, 2007 Contrast limitation due to pump noise in an optical parametric chirped pulse amplification system. Optics Communications, 273 2 May 2007), 510 514 , 0030-4018

38. H. Sayinc, U. Buenting, D. Wandt, J. Neumann, D. Kracht, 2009 Ultrafast high power Yb:KLuW regenerative amplifier. Opt. Express, 17 17 August 2009), 15068 15071 , 1094-4087

39. M. Siebold, M. Hornung, J. Hein, G. Paunescu, R. Sauerbrey, T. Bergmann, G. Hollemann, 2004 A high-average-power diode-pumped Nd:YVO4 regenerative laser amplifier for picosecond-pulses. Applied Physics B, 78 3-4 (February 2004), 287-290, 0946-2171

40. M. Siebold, J. Hein, M. Hornung, S. Podleska, M. Kaluza, S. Bock, R. Sauerbrey, 2008 Diode-pumped lasers for ultra-high peak power. Appl. Phys. B, 90 3-4 (March 2008), 431 437 , 0946-2171

41. J. Speiser, A. Giesen, 2008 Scaling of thin disk pulse amplifiers. Proc. SPIE, 6871 February 2008), 68710J, 0277-786X

42. D. Strickland, G. Mourou, 1985 Compression of amplified chirped optical pulses. Opt. Comm., 56 3 March 1985), 219 221 , 0030-4018

43. O. Svelto, 1998 Principles of Lasers, Plenum Press, 0-30645-748-2York

44. D. Tang, S. Ng, L. Qin, X. Meng, 2003 Deterministic chaos in a diode-pumped Nd:YAG laser passively Q switched by a Cr 4+ YAG crystal. Opt. Lett., 28 5 March 2003), 325-327, 0146-9592

45. S. Valling, T. Fordell, A. Lindberg, 2005 Experimental and numerical intensity time series of an optically injected solid state laser. Opt. Commun., 254 4-6 (October 2005), 282 289 -, 0030-4018

46. B. Walker, C. Toth, D. Fittinghoff, T. Guo, D. Kim, C. Rose-Petruck, J. Squier, K. Yamakawa, K. Wilson, B. Barty, 1999 A 50 EW/cm2 Ti:sapphire laser system for studying relativistic light-matter interactions. Opt. Express, 5 10 November 1999), 196-202, 1094-4087

CITATION

Mikhail Grishin and Andrejus Michailovas (2010). Dynamics of Continuously Pumped Solid-State Regenerative Amplifiers, Advances in Solid State Lasers Development and Applications, Mikhail Grishin (Ed.), ISBN: 978-953-7619-80-0, InTech, DOI: 10.5772/7953.

CHAPTER 6

Algorithm Based Comparison between the Integral Method and Harmonic Analysis of the Timing Jitter of Diode-Based and Solid-State Pulsed Laser Sources

N.K. Metzger[1], C.-R. Su[2], T.J. Edwards[2] and C.T.A. Brown[2]

[1]SUPA School of Engineering and Physical Sciences, Photonics & Quantum Sciences, Heriot Watt University, Edinburgh EH14 4AS, UK
[2]SUPA School of Physics and Astronomy, University of St Andrews, North Haugh, St Andrews, Fife KY16 9SS, UK

ABSTRACT

A comparison between two methods of timing jitter calculation is presented. The integral method utilizes spectral area of the single side-band (SSB) phase noise spectrum to calculate root mean square (rms) timing jitter. In contrast the harmonic analysis exploits the uppermost noise power in high harmonics to retrieve timing fluctuation. The results obtained show that a consistent timing jitter of 1.2 ps is found by the integral method and harmonic analysis in gain-switched laser diodes with an external cavity scheme. A comparison of the two approaches in noise measurement of a diode-pumped Yb:KY(WO$_4$)$_2$ passively mode-locked laser is also shown in which both techniques give 2 ps rms timing jitter.

INTRODUCTION

Actively and passively mode-locked lasers are ideal candidates for the generation of coherent, stable and highly periodical pulse trains. They have been the subject of intense investigation as they can be used in many applications, including high-speed optical communications, all-optical signal-processing, optical sampling and clock distribution [1]. Among these applications, some require not only high peak power and short pulse operation, but also the smallest possible timing jitter, as the fluctuation of the time interval between pulses degrades the quality of the expected system performance.

A broad bandwidth oscillator can detect timing fluctuations by monitoring the beat frequency between a modulated signal and low-jitter electrical trigger signal [2]. This enables one to easily obtain the exact timing fluctuation of an unknown optical source. Although this is an accurate method to calculate timing jitter, the equipment requirement of a broad bandwidth and a trigger-dependent source limits the practicality of such spectral measurements. These drawbacks can be overcome with the combination of a fast photodetector and an electronic spectrum analyzer. The available bandwidth of the spectrum analyzer not only facilitates measurements, but also provides important insight into sources of both correlated and uncorrelated timing jitter [2] and [3].

By considering the noise sideband of the power spectrum, phase noise can be distinguished from amplitude fluctuation. Whilst phase noise rises quadratically with harmonic order, amplitude fluctuation remains order independent across the full frequency spectrum [2]. Both of these effects contribute to pedestals or broad noise sidebands that prevent a clean RF signal. The single side-band (SSB) phase noise, which is identified by the carrier per resolution bandwidth, reveals information about the timing jitter. Following von der Linde's work [2], the rms timing fluctuation at a given carrier frequency f_R can be obtained from a spectral integration of noise if the rms amplitude noise remains small.

When using this method the integration boundaries need to be chosen carefully [4], [5], [6], [7], [8], [9], [10], [11] and [12] to obtain high measurement accuracy. To solve this problem, an approach using a simplified theoretical model has been developed [2]. The harmonic approach adjusts the integration by utilizing the uppermost noise power to identify the timing jitter in higher harmonic orders. This has been verified as a valid solution [9] and [10], yet it remains relatively little used compared to the integration method discussed above. This is because the accuracy of harmonic analysis is greatly restricted by the highest harmonic order obtained; the higher the harmonic order, the more precise the timing jitter.

To date, a thorough comparison of both methods has yet to be conducted, despite Ng et al. [11] and Yoshida [12] confirming the consistent outcome of these two approaches in their own system. This publication presents two studies of the rms timing jitter calculated by the integral method as well as the harmonic analysis approach in gain-switched semiconductor laser diodes. The calculated timing jitter of 1.1 ps and 1.25 ps obtained by harmonic analysis and integral method respectively in this work, proved comparable to the 1.5 ps jitter in a Fabry–Perot gain-switched semiconductor laser diode with optical feedback [13] and [14]. These two algorithms were then applied to an Yb:KYW passive mode-locked lasers and yielded a free-running jitter time of 2.05 ps and 1.95 ps respectively. A similar agreement of results was obtained with an Yb:Eb:glass ultrafast laser [15].

The outcomes of this study validates the consistent measurement of timing jitter by both the harmonic analysis and integral method when tested theoretically and experimentally in mode-locked and gain-switched lasers.

BACKGROUND TO JITTER MEASUREMENTS AND ALGORITHM DEVELOPMENT

The well-developed theory by von der Linde has been used to calculate rms timing jitter in spectral measurements [2]. This work analyzed the noise behavior in the power spectrum and found that noise varies with increasing harmonic orders. While amplitude noise remains a frequency-independent trend, phase noise increases quadratically and further becomes the main source of noise for high harmonics in RFSA. Phase noise is thus observed to have the largest contribution with respect to rms timing jitter. To compute timing jitter, there are two approaches advocated. The first approach, the integral method, uses the integration of the entire SSB phase noise to calculate the rms timing jitter. This approach assumes that the amplitude fluctuation is negligible in affecting the power spectrum. The second, harmonic analysis, is a simplified version of the integral method. This utilizes the whole power spectrum and then retrieves the uppermost noise power and full width at half maximum (FWHM) noise bandwidths in high harmonic orders for the calculation of rms timing jitter. Both methods have been validated to be correct theoretically and experimentally [3], [9], [11], [12], [16] and [17].

In order to compare rms timing jitter efficiently, these two methods were implemented in Matlab. The content of these programs for harmonic analysis and the integral method will be discussed in the following sections.

Algorithm for Harmonic Analysis
Before calculating rms timing jitter, harmonic analysis requires information from the RFSA trace. A fluctuation-free pulse train is seen to have a delta RF linewidth. However, once the pulse encounters phase noise, an undefined phase relation between each pulse will result in the broadened linewidth RFSA trace shown in Fig. 1. After the red crosses (*PB*), the power spectrum encounters noise interruption.

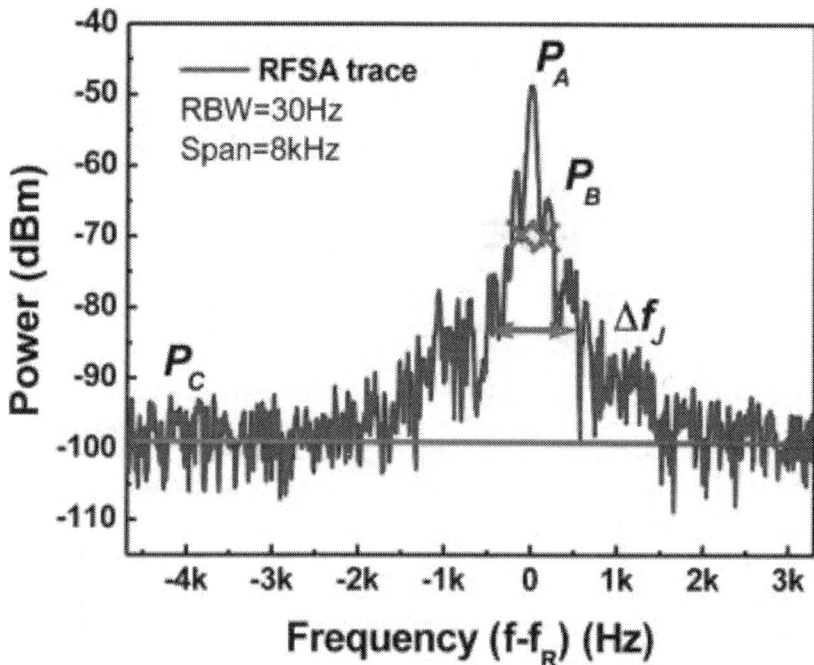

Figure 1. The RFSA trace obtained with Matlab (RBW=30 Hz). The RFSA trace shows large noise interference until the trace reaches the noise floor. *PA* marks the peak of the RFSA trace. *PB* represents the power of the maximum noise level and *PC* the averaged noise floor while ΔfJ is the FWHM level between *Pc* and *PB*. (For interpretation of the references to color in this figure, the reader is referred to the web version of this article.)

In Fig. 1, *PA* denotes the peak of the RFSA trace. *PB* and *PC* represent the power of the maximum noise level and the averaged noise floor respectively.

The rms timing jitter can be estimated [2] with Eq. (1), given prior knowledge of parameters *PA*, *PB*, *PC*, $\Delta fres$ resolution bandwidth (RBW), ΔfJ (the FWHM of noise bands), round trip time *T*, and the harmonic order *n* of the measured RFSA data in Eq.(1).

$$\Delta t = T(2\pi n)\left[\left(\frac{P_B}{P_A}\right)_n \frac{\Delta f_J}{\Delta f_{res}}\right]^{1/2}$$

(1)

where Δt symbolizes the rms timing jitter. *PB* the uppermost noise level (red crosses in Fig. 1) is determined automatically as illustrated in Fig. 2.

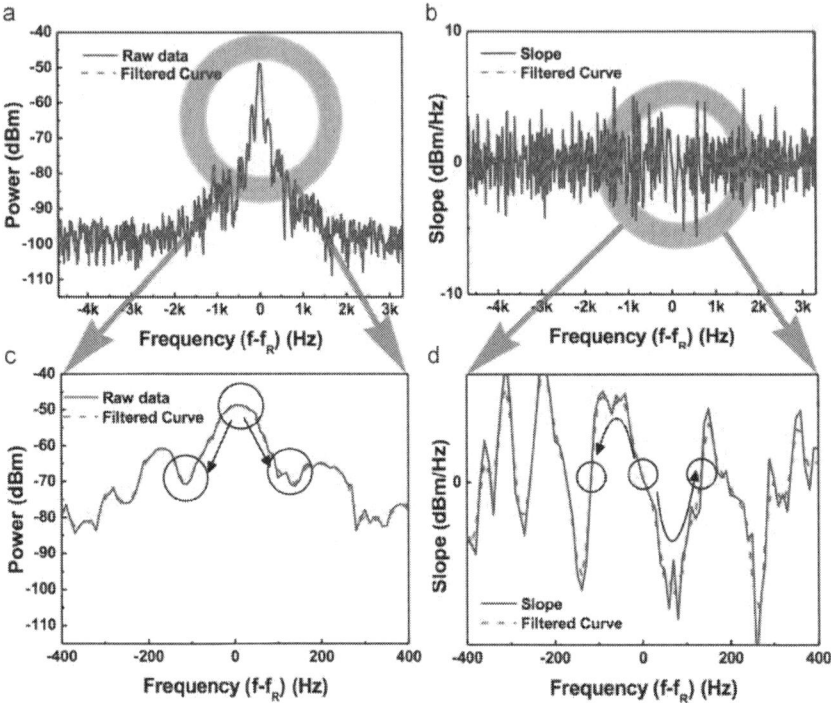

Figure 2. The illustration of harmonic analysis (a) the raw RFSA trace (blue curve) and the filtered curve (red dashed line) with a RBW of 30 Hz. (b) The slope of the RFSA data (blue curve) and its filtered curve (red dashed). (c) The 800 Hz span zoomed-in version of (a). The black arrows indicate the direction of the search of the algorithm. (d) 800 Hz span zoomed-in version of (b). The black arrows indicate the direction of the search of the algorithm. (For interpretation of the references to color in this figure legend, the reader is referred to the web version of this article.)

Initially, the unprocessed RFSA data is convoluted with a moving average filter generated by the analysis program to remove spurious aliasing ripples. From this filtered curve Fig. 2(a), the algorithm is able to extract its slope. The slope of the raw RFSA data and its filtered curve slope are both depicted in Fig. 2(b) by the blue curve and red dotted curve, respectively. Looking back at the original data in Fig. 2(a), it can be seen that the uppermost phase noise has a local minima, shown in the expanded (c). Furthermore, the local maximum or minimum points in Fig. 2(c) will have a value of approximately zero at the same frequencies in (d). It is not surprising that the extreme values usually result from the transition in slope. Therefore, to find the uppermost noise, the program can rely on the transition point of the slope. In other words, values where there are zero crossingsFig. 2(d) correspond to the extreme values of (c).

Similarly *PA* and *PC* can both be obtained from a standard maximum and minimum search function where *PC* was taken as the average noise floor. The points (fJ_1 and fJ_2) both have powers of $(PB+PC)/2$ in the power spectrum, where the corresponding frequency distance between the two points determines the FWHM of the noise bandwidths $\Delta fJ = fJ_1 - fJ_2$. Roundtrip time *T*, harmonic order *n*, and RBW can be retrieved from the RFSA trace. This method assumes no correlations between the intensity and phase noise of the pulses [2]. When timing-jitter fluctuations between pulses are uncorrelated in time a Lorentzian shaped RFSA trace is obtained. Correlations between timing fluctuations tend to produce traces that are Gaussian in shape [6]. In the first case the accuracy of the prediction of the jitter will suffer however the algorithm will still give a fast qualitative prediction as shown in experimental section.

Amplitude fluctuations

Although amplitude noise remains a small value for most frequencies, it nevertheless influences the power spectrum regarding lower harmonic orders [2]. Zero-order noise has especially been found to be influenced most by energy instability. This behavior is described below in Eq. (2)

$$\Delta E/E = [(P_B/P_A)_0 \Delta f_A/\Delta f_{res}]^{1/2}$$

$$(2)$$

where $\Delta E/E$ is the amplitude fluctuation, *PB* is the maximum of noise level, *PA* is the peak power, ΔfA is the FWHM of the zero order noise bands and $\Delta fres$ is the resolution bandwidth. It can be seen that the equation is very similar to Eq. (1), except that the harmonic order is zero. Therefore the Matlab based harmonic analysis algorithm can be applied to evaluate energy fluctuation when the carrier frequency is set at DC.

Integral Method

In a second approach the integral method was used to automatically determine the noise level. When using the integral method to calculate the timing jitter, it is necessary to obtain the SSB phase noise spectrum. The SSB phase noise spectrum, $L(f)$, is defined as the ratio of noise in a 1 Hz resolution bandwidth at a specified frequency offset f to the oscillator signal amplitude at carrier frequency fn. Eq. (3) illustrates this concept [16]

$$L(f) = \log_{10}\left[\frac{P_F(f)}{P_A \Delta f_{res}}\right]$$

(3)

where $PF(f)$ is the SSB noise spectral density, PA is the carrier power, and Δf_{res}resolution bandwidth (RBW). For higher harmonic orders, $PF(f)$ will be dominated by the SSB phase noise spectral density $PJ(f)$ [16]. Where $L(f)$ is obtained from the RFSA trace in units of dBc/Hz. The spectral area of the power spectrum can be directly related to the timing fluctuation by Eq. (4)

$$t_j = \frac{1}{2\pi f_n}\sqrt{\int_{f_n+f_{min}}^{f_n+f_{max}} L(f)df + \int_{f_n-f_{max}}^{f_n-f_{min}} L(f)df}$$

(4)

where fn is the carrier frequency at harmonic order n, $L(f)$ is the SSB phase noise spectrum, and tj is the rms timing jitter. In the equation, $fmax$ and $fmin$ are the upper and lower boundaries for integration.

Because of its simple numerical formula, the integral method is most commonly used for calculating timing jitter. An algorithm for the integration method can be implemented by simply performing integration via the trapezoidal method.

Despite being a convenient way to determine timing jitter, care has to be taken when defining the integration boundaries $fmax$ (upper boundary) and $fmin$ (lower boundary) [2].

In this work the lower integration boundary (*fmin*) was determined through the aid of the harmonic algorithm. Fig. 3(a) shows the 14th order SSB phase noise $L(f)$ in the RF spectrum. The red cross represents *PB* the power of the maximum noise level as used for the harmonic analysis. In Fig. 3(b) the normalized $L(f)$ from harmonic orders 1–14 is shown.

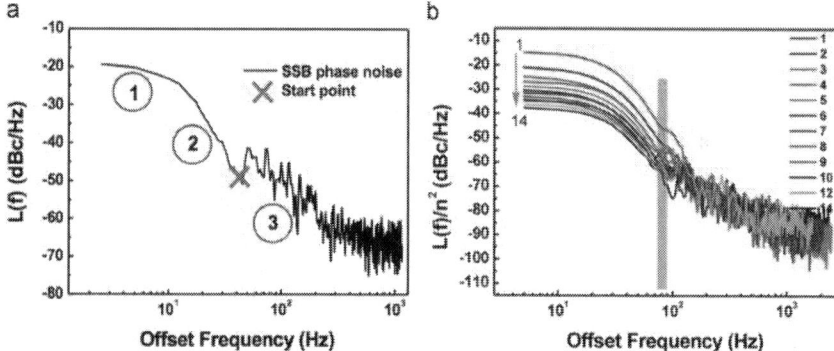

Figure 3. (a) The SSB phase noise spectrum (acquired with a RBW of 80 Hz) of harmonic order $n=14$ with three different slopes; (b) the normalized SSB phase noise depicted with varied harmonic orders from $n=1$ to 14. With *fmin* being within offset frequencies 80–90 Hz as highlighted in red. (For interpretation of the references to color in this figure legend, the reader is referred to the web version of this article.)

The curve in Fig. 3(a) can be divided into three regions. The first section from 5 Hz to 20 Hz displays a white plateau as marked 1. The second section ranging from 20 Hz to 40 Hz shows accelerated degradation. Beyond 40 Hz, the third section shows a decaying trend of 20 dBc/Hz per decade due to phase noise marked as 3. Fig. 3(b) demonstrates the respective normalized SSB phase noise with varying order *n*. The lower integration boundary (*fmin*) are within offset frequencies 80–90 Hz as highlighted in Fig. 3(b). After this the normalized SSB phase noise scales and aligns each other accordingly to their respective order number *n* from 1 to 14.

These lower boundaries were determined by the harmonic analysis algorithm as indicated by the red cross in Fig. 3(a). According to the algorithm's search result, the pure signal and phase noise can be separated, thus demonstrating the viability of the Matlab algorithm to identify phase noise.

The upper integration boundary *fmax* has an approximate limit. Typically *fmax* is half the span [11] and [18] or the point where phase noise hits the thermal noise floor [4] and [19]. In the analysis presented here the upper integration boundary was selected to be at the intercept where the noise floor meets the phase noise, to be analogous with the harmonic analysis and the slope is not at 20 dBc/Hz per decade. In the following example this is illustrated: the integral algorithm employed *fmin*=0.5 kHz, found by the harmonic algorithm as a lower integration boundary and the upper integration boundary was *fmax*=10 kHz as the intercept with the noise floor. The noise floor therefore determines the upper boundary and hence these fluctuations are not accounted for in the integral. This aids to reduce the contribution of amplitude fluctuations to the analysis under the assumption that the timing jitter within the integral boarders is correlated and amplitude fluctuations are small and do not affect the phase [2]. Fig. 4 demonstrates the resulting integration boundaries.

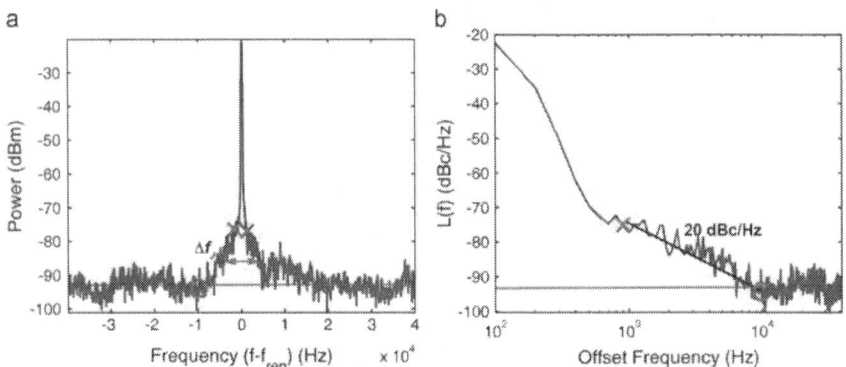

Figure 4. (a) Harmonic analysis: The RFSA trace (with a RBW of 100 Hz) shows *PB* (red crosses) and *PC* the average noise floor as a red line as determined by the harmonic algorithm. Addtionally the Δ*fJ* (the FWHM of the noise band) is indicated with the double arrows. The red circles are used as *fmax* by the integral analysis of the left hand SSB. (b) Integral analysis: *fmin* (red cross) is determined by the harmonic analysis. The red circle indicates *fmax* where the noise band intercepts the noise floor. The characteristic 20 dBc/Hz slope is observed between *fmin* and *fmax*. (For interpretation of the references to color in this figure legend, the reader is referred to the web version of this article.)

In the analysis shown in Fig. 4, a jitter of 2.85 ps was obtained compared to a jitter value of 2.59 ps obtained for the same dataset using the

Harmonic analysis. To check the robustness of our approach several measurements were conducted from which a precision of ±6% was estimated.

Integration of the entire area from *fmin* to half of the measurement span as used in Refs.[18] and [20] results in an increase of the calculated jitter by about 15%, which is similar to[21]. Thus it can be seen that the spectral area related to phase noise is affected by the correct choice of *fmax* outside the thermal noise floor. The average noise floor (red line inFig. 4(a)) was found to be −96.42 dBm whereas *fmax* (red circle in Fig. 4(a)) was determined to −93.45 dBm by the integral analysis. Both values are reasonably close to each other and give confidence in the presented approach to determine the integration boundary. In the following work, *fmin* was set to be equal to that used in the harmonic analysis to enable the comparison between the two approaches.

With these algorithms, we are now well placed to analyze different laser systems and compare their jitter values.

APPLICATION TO PICOSECOND PULSES FROM A GAIN-SWITCHED EXTERNAL CAVITY DIODE LASER

With the developed algorithm, we assessed the performance of a commercial GaAs tapered diode lasers from m2k-laser (TAL-1060-2000). The laser has a reflectivity 1% at the output facet and anti-reflection coating of 0.01% at the rear facet and was arranged in an external cavity geometry [22] and [23] as depicted in Fig. 5(a). The laser was operated with a DC offset bias and a supplementary RF-modulated injection current from a signal generator. An electronic amplifier was incorporated to provide an RF power of 35 dBm. The output optical pulses centered at 1060 nm were detected using a fast InGaAs photodiode, which had a time response full-width at half-maximum (FWHM) of 12.5 ps. Care has to be taken not to oversaturate the sensitive detector as permanent damage might occur. For the experiments conducted here an average power of 0.5–0.6 mW was incident on the detector element. This was below the maximum value of 1 mW quoted by the manufacturer for save operation. By blocking the beam into the detector input it was ensured that the acquired RFSA trace was above the detector and RF analyzer noise floor before each measurement. The pulse train was captured using a 22 GHz Radio Frequency Signal Analyzer that was computer interfaced.

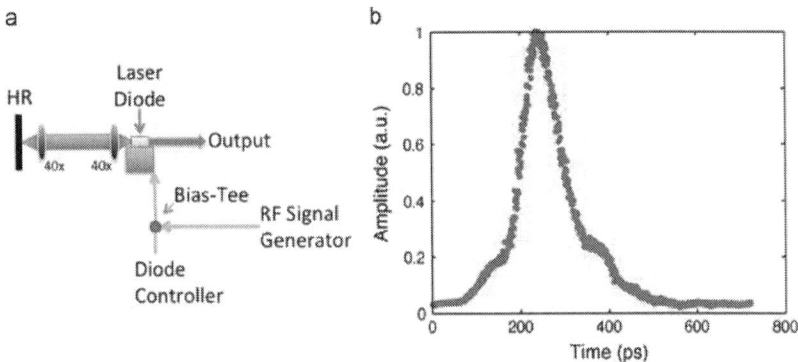

Figure 5. (a) Layout of the gain-switched laser diode. The external cavity is formed between a HR high reflected mirror and the diode facet. Two 40× microscope objectives are used to collimate the beam and focus it into the laser diode. A laser diode driver that provided the DC offset in combination with a signal generator via a bias-tee operated the diode. (b) Example pulse obtained from the gain-switched laser diode with pulse duration of about 80 ps at a repetition rate of 0.64 GHz (~23 cm external cavity length).

The maximum output power is found to be 115 mW when operated at an injection current of 200 mA. The cavity length is chosen to be 23.2 cm give a pulse repetition frequency of 651.1 MHz. The best pulse quality requires a stable signal without any noise induced by self-pulsation. The RFSA trace is found to be Lorentzian shape when the DC current is maintained at 42 mA, and the RF frequency is set at 640.67 MHz. These results correspond to those obtained in similar gain-switched schemes [13]. With the parameters chosen above, the RFSA trace can be seen up to the 7th harmonic order in the system.

The jitter algorithm was applied to all seven harmonic orders. The harmonic approach and integral method in the RF spectrum are depicted in Fig. 6(a) and (b), respectively.

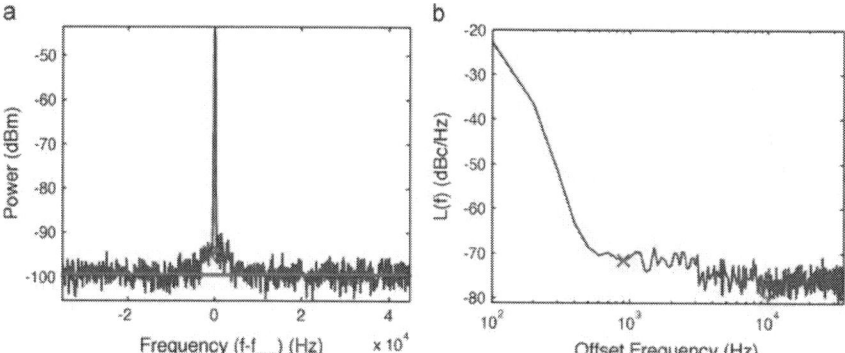

Figure 6. (a) The RFSA trace of the 7th order in gain-switched laser diode with a RBW of 100 Hz and span of 80 kHz (b) the SSB phase noise spectrum of 7th order with the lower integration $fmax$=10 kHz (circle) and the upper boundary $fmin$=900 Hz (cross). (For interpretation of the references to color in this figure, the reader is referred to the web version of this article.)

In Fig. 6(a), the harmonic approach establishes the red starred points as the uppermost powers used to calculate timing jitter. From Fig. 6(b), the spectral area is calculated via integration from $fmin$=0.9 kHz to $fmax$=10 kHz and is further converted to timing jitter by the integral method. Increasing of the integration border to the full acquisition span $fmax$=40 kHz results in an increase of the timing jitter by 13% approximately 3% relative to the increased integration bandwidth indicating that the major contribution to the jitter is in the lower frequency components. Relaxation oscillations for modulated lasers can contribute to the timing jitter but are estimated to be higher than the jitter frequency range of approximately 1–10 kHz considered here. Fig. 7(a) shows that the total phase noise determined by the harmonic and integral method is consistent. The two results both follow the linear tendency of $\phi(n)/\phi(1)=n$ from orders 1 to 7. Consequently, a frequency-independent jitter is obtained from the two algorithms in Fig. 7(b).

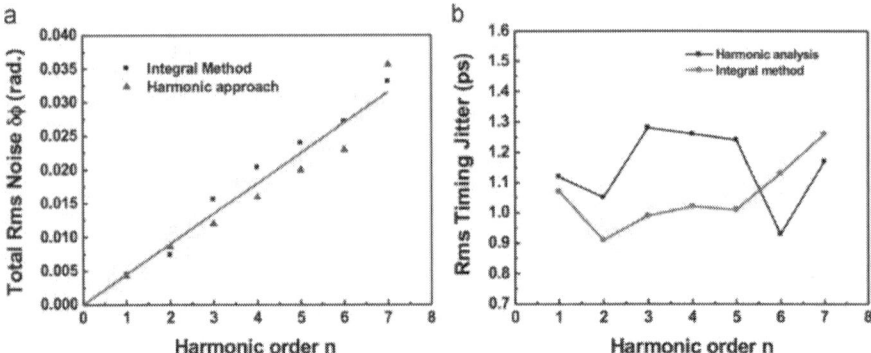

Figure 7. (a) The rms phase fluctuation determined by the harmonic and integral method and plotted with the theoretical prediction line. (b) The measurement of timing jitter by the two algorithms.

From Fig. 7(b), the average jitter [5] is determined to 1.2 ± 0.2 ps and 1.2 ± 0.2 ps for the harmonic and integral method respectively. This corresponds well to the value (1.5 ps) obtained in [13] where the single contact Fabry–Perot gain-switched semiconductor laser diode is operated at twice the DC threshold current with an external cavity scheme. The reduced jitter, which differs from the typical value of the gain-switched edge emitting laser diode (>1.5 ps experimentally and >3.5 ps theoretically [24]), is due to optical feedback, which has been verified to reduce timing jitter greatly [25]. This is because the high coherence of reflected photons suppresses the spontaneous emission of laser diodes [24]. In summary, both the harmonic approach and the integral method give consistent and accurate calculations of timing jitter for this laser system.

APPLICATION TO FEMTOSECOND PULSES FROM AN MODE-LOCKED YB:KYW LASER

We next applied the algorithm to passively mode-locked pulses from a solid state laser. The pulse source used was a diode-pumped Yb:KYW laser similar to the systems developed in [26] and [27]. The laser was passively mode-locked using a semiconductor saturable absorber mirror and produced of 138 fs duration in the 1035–1040 nm wavelength range with a pulse repetition frequency of 161 MHz at an average output power of 80 mW. The corresponding spectral width for the pulses was 8.5 nm, which implied a time-bandwidth product of 0.33.

The schematic and a measured intensity autocorrelation trace of the mode-locked Yb:KYW laser are shown in Fig. 8.

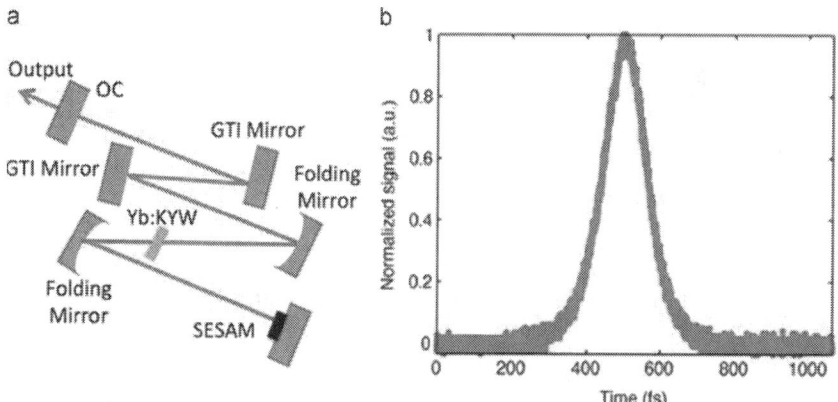

Figure 8. (a) A schematic of the asymmetric z-fold Yb:KYW cavity. The folding mirrors are spherically curved with radii of curvature of 75 mm. SESAM is a semiconductor saturable absorber mirror (A=2% and ΔR=1.2%), while OC is a 3.2% output coupler. Both GTI mirrors provide a single pass dispersion of −910 fs². (b) A representative autocorrelation trace of the pulses obtained from the mode-locked Yb:KYW laser with a pulse duration of 138 fs.

The output beam is coupled into the same measurement setup as in the previous section via an optical isolator to prevent feedback. The highest harmonic order is found to be 22. In harmonic orders higher than 22, the phase noise interferes with the RFSA trace so strongly that the timing jitter cannot be distinguished.

A high RBW of 80 Hz and span of 20 kHz is chosen to prevent amplitude fluctuation from entering low harmonics. In Fig. 9 the RFSA trace and the resulting SSB trace for the 7th order are shown.

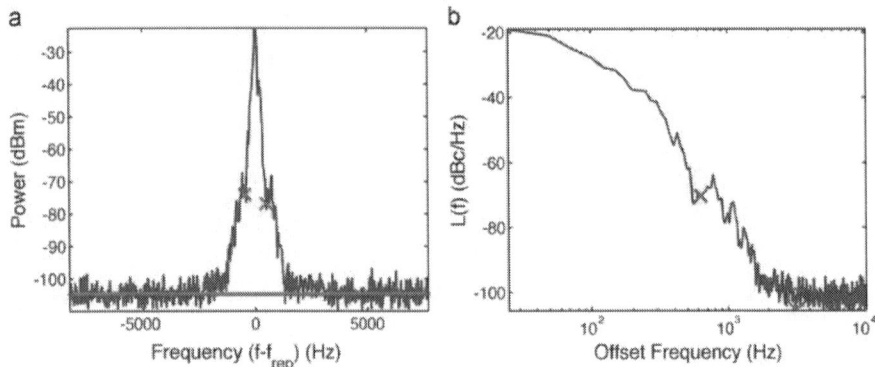

Figure 9. (a) The RFSA trace for the 7th order with a RBW of 80 Hz, and (b) its SSB phase noise with the lower integration *fmax*=3.2 kHz (circle) and the upper boundary *fmin*=625 Hz (cross).

In Fig. 9(b), the noise band has a gradually descending slope in the SSB phase noise spectrum and is found that the noise skirt hits the noise floor at an offset frequency of approximately 3 kHz. Hence, *fmax*=3.1 kHz is selected as a suitable upper boundary and *fmin*=625 Hz is selected as the lower boundary.

Fig. 10(a) shows the evaluation of timing jitter when the amplitude noise is taken to be negligible. Fig. 10(b) illustrates the total rms phase noise with varied harmonic orders.

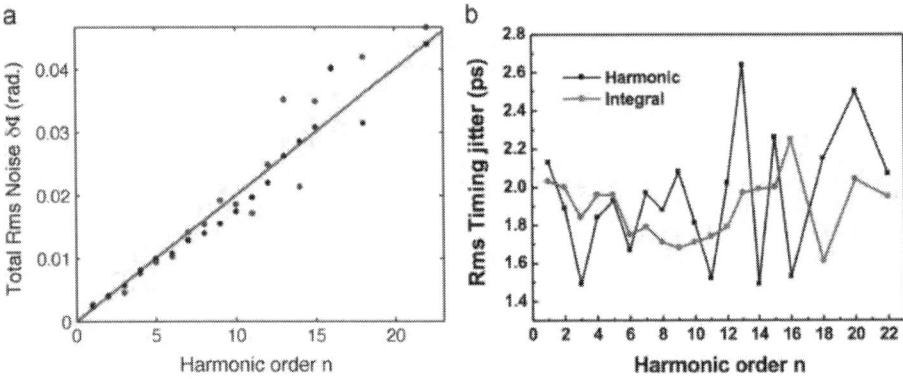

Figure 10. The timing jitter measurements of a Yb:KYW mode-locked laser (a) the total phase noise fluctuation and (b) the timing jitter of different harmonic orders calculated by the two algorithms.

In Fig. 10(a), the linear trend also obeys the predicted theoretical model detailed previously. In Fig. 10(b), the consistency of harmonic approach and integral method is demonstrated up to the 12th harmonic order. In higher harmonic orders, the phase noise interferes with the RFSA trace more strongly so that the timing jitter cannot be distinguished accurately and the results start to fluctuate. The algorithms determine an average timing jitter [5] of 2.0±0.6 ps and 1.9±0.3 ps for the harmonic and integral method respectively for the different orders. This is a comparable value to the 2.5 ps obtained in a similar soliton mode-locked system [15]. Timing phase noise scales inversely with the intracavity pulse energy and linearly with the square of pulse duration [28]. The discrepancy of 0.5 ps can be explained from the interaction between the lower pulse energy (0.5 nJ) and shorter pulse duration (138 fs) compared to the pulse energy (1.03 nJ) and pulse duration (255 fs) obtained in [15]. In conclusion, harmonic and integral method found an average timing jitter of approximately 2 ps, a value that is much larger than that obtained in active mode-locked systems [14]. To diminish the free-running timing jitter to a femtosecond jitter regime, feedback timing stabilization such as a phase-locked loop (PLL) system can be employed in the same mode-locked system scheme [29].

CONCLUSION

This publication presented the first direct comparison between the harmonic analysis and the integral method of characterizing timing jitter. The study of both approaches used the theoretical framework developed by von der Linde [2] that has been widely used in RF measurement, and investigated using an automated Matlab program [30]. The algorithm results show that both the harmonic approach and the integral method correspond to the theory appropriately and are reliable in characterizing rms timing jitter. Noise estimation in gain-switched laser diode and Yb:KYW passive mode-locked solid state laser are used to thoroughly compare the two methods.

The applied method relies on direct detection of the pulse train and therefore the dynamic range of the detector limits phase noise measurement of the RF harmonic. These limitations are overcome by employing the optical cross-correlation method [31]. In contrast to the approach applied in this publication it is an all-optical timing jitter characterization method that enables extremely high timing resolution. The optical cross-correlation method has recently been employed for ultralow timing jitter measurement [32] in solid state lasers and also shows good agreement with the integral method when applied to mode-locked laser diodes [33].

Nevertheless the outcomes not only demonstrate a consistent relationship between the two methods but also prove their accuracy in evaluating rms timing jitter, as the values have been confirmed experimentally and theoretically in the same system setup.

ACKNOWLEDGMENTS

This work was supported by the European Metrological Research Programme EMR Punder IND 14. The EMRP is jointly funded by the EMRP participating countries within EURAMET and the European Union. N.K. Metzger acknowledges support from the EPSRC Centre for Innovative Manufacturing in Laser-based Production Processes, funded through EPSRC grant EP/K030884/1.

REFERENCES

1. E.A. Avrutin, J.H. Marsh, E.L. Portnoi, Monolithic and multi-GigaHertz modelocked semiconductor lasers: constructions, experiments, models and applications, IEEE Proc.—Optoelectron. 147 (2000) 251–278.
2. D. von der Linde, Characterization of the noise in continuously operating mode-locked lasers, Appl. Phys. B 39 (1986) 201–217.
3. D.A. Leep, D.A. Holm, Spectral measurement of timing jitter in gain-switched semiconductor lasers, Appl. Phys. Lett. 60 (1992) 2451–2453.

4. J.P. Tourrenc, A. Akrout, K. Merghem, A. Martinez, F. Lelarge, A. Shen, G. H. Duan, A. Ramdane, Experimental investigation of the timing jitter in selfpulsating quantum-dash lasers operating at 1.55 mm, Opt. Exp. 16 (2008) 17706–17713.

5. C.Y. Lin, F. Grillot, Y. Li, R. Raghunathan, L.F. Lester, Microwave characterization and stabilization of timing jitter in a quantum-dot passively mode-locked laser via external optical feedback, IEEE J. Sel. Top. Quantum Electron. 17 (2011) 1311–1317.

6. D. Eliyahu, R.A. Salvatore, A. Yariv, Effect of noise on the power spectrum of passively mode-locked lasers, J. Opt. Soc. Am. B 14 (1997) 167–174.

7. B. Zhu, I.H. White, K.A. Williams, M.R.T. Tan, R.P. Schneider Jr, S.W. Corzine, S. Y. Wang, Ultralow timing jitter picosecond pulse generation from electrically gain-switched oxidized vertical-cavity surface-emitting lasers, IEEE Photonics Technol. Lett. 9 (1997) 1307–1309.

8. M. Schell, W. Utz, D. Huhse, J. Kässner, D. Bimberg, Low jitter single-mode pulse generation by a self-seeded, gain-switched fabry-perot semiconductorlaser, Appl. Phys. Lett. 65 (1994) 3045–3047.

9. M. Jinno, Correlated and uncorrelated timing jitter in gain-switched laserdiodes, IEEE Photonics Technol. Lett. 5 (1993) 1140–1143.

10. A. Martinez, S. Yamashita, Multi-gigahertz repetition rate passively modelocked fiber lasers using carbon nanotubes, Opt. Exp. 19 (2011) 6155–6163.

11. W. Ng, Y.M. So, R. Stephens, D. Persechini, Characterization of the jitter in a mode-locked Er-fiber laser and its application in photonic sampling for analog-to-digital conversion at 10 Gsample/s, J. Lightwave Technol. 22 (2004) 1953–1961.

12. E. Yoshida, M. Nakazawa, Measurement of the timing jitter and pulse energy fluctuation of a PLL regeneratively mode-locked fiber laser, IEEE Photonics Technol. Lett. 11 (1999) 548–550.

13. K.A. Williams, I.H. White, D. Burns, W. Sibbett, Jitter reduction through feedback for picosecond pulsed InGaAsP lasers, IEEE J. Quantum Electron. 32 (1996) 1988–1994.

14. D.J. Derickson, P.A. Morton, J.E. Bowers, R.L. Thornton, Comparison of timing jitter in external and monolithic cavity mode-locked semiconductor-lasers, Appl. Phys. Lett. 59 (1991) 3372–3374.

15. G.J. Spühler, L. Krainer, E. Innerhofer, R. Paschotta, K.J. Weingarten, U. Keller, Soliton mode-locked Er:Yb:glass laser, Opt. Lett. 30 (2005) 263–265.

16. K.K. Gupta, D. Novak, H.F. Liu, Noise characterization of a regeneratively modelocked fiber ring laser, IEEE J. Quantum Electron. 36 (2000) 70–78.

17. M.J.W. Rodwell, D.M. Bloom, K.J. Weingarten, Subpicosecond laser timing stabilization, IEEE J. Quantum Electron. 25 (1989) 817–827.

18. U. Keller, K.D. Li, M.J.W. Rodwell, D.M. Bloom, Noise characterization of femtosecond fiber raman soliton lasers, IEEE J. Quantum Electron. 25 (1989) 280–288.

19. G. Serafino, P. Ghelfi, P. Pérez-Millán, G.E. Villanueva, J. Palací, J.L. Cruz, A. Bogoni, Phase and amplitude stability of EHF-band radar carriers generated from an active mode-locked laser, J. Lightwave Technol. 29 (2011) 3551–3559.

20. A. Finch, X. Zhu, P.N. Kean, W. Sibbett, Noise characterization of mode-locked color-center laser sources, IEEE J. Quantum Electron. 26 (1990) 1115–1123.

21. M.J.R. Heck, E.J. Salumbides, A. Renault, E.A.J.M. Bente, Y.-S. Oei, M.K. Smit, R. van Veldhoven, R. Nötzel, K.S.E. Eikema, W. Ubachs, Analysis of hybrid mode-locking of two-section quantum dot lasers operating at 1.5 μm, Opt. Exp. 17 (2009) 18036.

22. P.J. Delfyett, L.T. Florez, N. Stoffel, T. Gmitter, N.C. Andreadakis, Y. Silberberg, J. P. Heritage, G.A. Alphonse, High-power ultrafast laser-diodes, IEEE J. Quantum Electron. 28 (1992) 2203–2219.

23. P.J. Delfyett, High-power ultrafast semiconductor-laser diodes, Ultrafast Pulse Gener. Spectrosc. 1861 (1993) 72–83.

24. M.R.H. Daza, C.A. Saloma, Jitter dynamics of a gainswitched semiconductor laser under self-feedback and external optical injection, IEEE J. Quantum Electron. 3 (2001) 254–264.

25. E.H. Bottcher, D. Bimberg, Detection of pulse to pulse timing jitter in periodically gain-switched semiconductor-lasers, Appl. Phys. Lett. 54 (1989) 1971–1973.

26. N.K. Metzger, W. Lubeigt, D. Burns, M. Griffith, L. Laycock, A.A. Lagatsky, C.T. A. Brown, W. Sibbett, Ultrashort-pulse laser with an intracavity phase shaping element, Opt. Exp. 18 (2010) 8123–8134.

27. A.A. Lagatsky, E.U. Rafailov, C.G. Leburn, C.T.A. Brown, N. Xiang, O. G. Okhotnikov, W. Sibbett, Highly efficient femtosecond Yb:KYW laser pumped by single narrow-stripe laser diode, IEEE Electron. Lett. 39 (2003) 1108–1110.

28. R. Paschotta, Noise of mode-locked lasers (Part II): timing jitter and other fluctuations, Appl. Phys. B 79 (2004) 163–173.

29. H. Tsuchida, Pulse timing stabilization of a mode-locked Cr:LiSAF laser, Opt. Lett. 24 (1999) 1641–1643.

30. N.K. Metzger, ⟨http://home.eps.hw.ac.uk/km359/commercial.html⟩ (August 2014).

31. L.A. Jiang, A. Leaf, M.E. Grein, H. Haus, E.P. Ippen, Noise of mode-locked semiconductor lasers, IEEE J. Sel. Top. Quantum Electron. 7 (2001) 159–167.

32. A.J. Benedick, J.G. Fujimoto, F.X. Kärtner, Optical flywheels with attosecond jitter, Nat. Photonics 6 (2012) 97–100.
33. T.K. Kim, Y. Song, K. Jung, C. Kim, H. Kim, C.H. Nam, J. Kim, Sub-100-as timing jitter optical pulse trains from mode-locked Er-fiber lasers, Opt. Lett. 36 (2011) 4443–4445.

CITATION

N.K. Metzger, C.-R. Su, T.J. Edwards, C.T.A. Brown, Algorithm based comparison between the integral method and harmonic analysis of the timing jitter of diode-based and solid-state pulsed laser sources, Optics Communications, Volume 341, 15 April 2015, Pages 7-14, ISSN 0030-4018, http://dx.doi.org/10.1016/j.optcom.2014.11.088.

CHAPTER 7

Simulating Spin Dynamics in Organic Solids under Heteronuclear Decoupling

Ilya Frantsuzov[1], Matthias Ernst[2], Steven P. Brown[3] and Paul Hodgkinson[1]

[1]Department of Chemistry, Durham University, South Road, Durham DH1 3LE, United Kingdom
[2]Laboratory of Physical Chemistry, ETH Zürich, Wolfgang-Pauli-Strasse 10, 8093 Zürich, Switzerland
[3]Department of Physics, University of Warwick, Coventry CV4 7AL, United Kingdom

ABSTRACT

Although considerable progress has been made in simulating the dynamics of multiple coupled nuclear spins, predicting the evolution of nuclear magnetisation in the presence of radio-frequency decoupling remains challenging. We use exact numerical simulations of the spin dynamics under simultaneous magic-angle spinning and RF decoupling to determine the extent to which numerical simulations can be used to predict the experimental performance of heteronuclear decoupling for the CW, TPPM and XiX sequences, using the methylene group of glycine as a model system. The signal decay times are shown to be strongly dependent on the largest spin order simulated. Unexpectedly large differences are observed between the dynamics with and without spin echoes. Qualitative trends are well reproduced by modestly sized spin system simulations, and the effects of finite spin-system size can, in favourable cases, be mitigated by extrapolation. Quantitative prediction of the behaviour in complex parameter spaces is found, however, to be very challenging, suggesting that there are significant limits to the role of numerical simulations in RF decoupling problems, even when specialist techniques, such as state-space restriction, are used.

INTRODUCTION

Effective decoupling of the ^{1}H nuclear spins is essential for achieving high resolution ^{13}C NMR spectra from typical organic molecules. Such heteronuclear decoupling is particularly difficult in the solid state due to the strong dipolar interactions between the different magnetic nuclei, which are not averaged out by molecular motion as they are in the solution state. While considerable progress has been made in developing approaches to decoupling and understanding how they work [1] and [2], there is not a comprehensive theory that allows decoupling performance to be predicted.

In principle, exact numerical simulation of nuclear spin systems [3], [4] and [5] ought to allow the prediction of decoupling performance. The multi-spin nature of the dipolar-coupled network in solid systems is not necessarily an obstacle; we have, for example, shown that simulations on modest numbers (9–10) of spins are sufficient to predict exactly the ^{1}H spin dynamics under magic-angle spinning (MAS) [6] and [7], and numerical solutions have proved invaluable in understanding how decoupling sequences work [8], [9], [10], [11], [12], [13], [14] and [15]. We show here, however, that predicting the performance of decoupling sequences, particularly in regions of interest around optimal conditions, would require very large numbers of spins to be included to obtain quantitative agreement with experiment. This number is larger than the practical limit for exact simulation, which is typically less than 12 spins (although this can be extended using artificial model geometries with additional symmetry [16]).

Recently a number of groups have demonstrated simulations of much larger numbers of coupled nuclear spins by restricting the size of the state space used for the simulations [17], [18], [19], [20] and [21]. Different researchers have used slightly different methods for restricting the evolution of the spin system to coherences below a certain order, but it is argued that the success of such calculations relies on the populations of higher spin orders (i.e. the number of correlated spins involved in a coherence) remaining relatively small [22]. This is clearly the case in solution-state NMR, where high spin order coherences relax relatively quickly, and some promising results have also been obtained for simulations of ^{1}H spin-diffusion in powder samples under MAS [23], [24] and [25]. It is not obvious, however, that state-space restriction is generally appropriate in the solid state, where very high spin orders can be observed amongst ^{1}H nuclei [26], [27] and [28], and so we also investigate the role of higher spin orders in heteronuclear decoupling.

METHODS

The decoupling performance is quantified experimentally by measuring the 'T_2' decay constant of the ^{13}C magnetisation, that is, the time constant for the decay of ^{13}C magnetisation where a 180° pulse is applied on ^{13}C at the mid-point of the decay period[29] and [30]. The spin-echo refocuses decay due to inhomogeneous effects, such as B_0inhomogeneity and magnetic susceptibility broadenings [31]. As a result, T_2 is much more sensitive to decoupling quality than T_2^* values obtained from measured linewidths; linewidths are often not particularly sensitive to changes in decoupling quality [30] and [32]. T_2 values continue to increase as the RF decoupling power is increased, tending towards the fundamental limit set by true T_2 relaxation [33], well after the limiting linewidth is observed. We use T_2^c here to refer to the *coherent* decay of ^{13}C magnetisation in the absence of a spin-echo, to distinguish it from true (incoherent) T_2 relaxation and the overall time constant for magnetisation decay, T_2^*, which includes inhomogeneous contributions associated with the sample and any instrumental factors. The mechanisms for the decay of ^{13}C magnetisation are different for T_2^c and T_2, but they both result in loss of the original coherence which cannot be readily refocussed. Decoherence is used here to refer to this magnetisation decay. In contrast to T_2 and T_2^*, T_2^c is not directly measurable, while both T_2 and T_2^c can, in principle, be directly observed in numerical simulations of magnetisation decay with or without a spin echo. It is important to note, however, that all the T_2 values used here are phenomenological quantities obtained by fitting experimental or simulated magnetisation decays to exponential functions. The absence of molecular tumbling in the solid state means that the magnetisation decays will generally be orientation dependent, and their powdered-averaged sum may not fit well to a single exponential.

Experimental

Experimental measurements of T_2' were performed on a polycrystalline sample of glycine-2-^{13}C,^{15}N (99% ^{13}C, 98% ^{15}N) purchased from CortecNet. The sample was confirmed to be α-glycine based on the ^{13}C carbonyl peak at 176.5 ppm, which is sensitive to polymorphic changes [34] and [35]. As expected from the stability range of this form, 5–500 K [36], no transformations were observed during experiments. For measurements of TPPM [37] and CW performance, the sample was packed into a Bruker 2.5 mm o.d. MAS rotor and data obtained at a ^{1}H Larmor frequency of $\nu_0^H = 600\,\text{MHz}$ and MAS frequency of $\nu_r = 12\,\text{kHz}$ on a Bruker Avance II+ console. The XiX [32] and [63] measurements were performed at $\nu_0^H = 500\,\text{MHz}$ on an InfinityPlus console using a Bruker 1.3 mm o.d. rotor spinning at $\nu_r = 25\,\text{kHz}$. ^{13}C magnetisation was created using cross-polarisation ramped on the ^{1}H RF power using either the centreband matching condition with a contact time of 1.2 ms, for $\nu_r = 12\,\text{kHz}$, or the $\nu_1^H - \nu_1^C = \nu_r$ sideband with a contact time of 2.7 ms for $\nu_r = 25\,\text{kHz}$. These represent typical conditions for acquiring ^{13}C spectra of natural abundance samples under TPPM and XiX decoupling. The ^{13}C magnetisation was then measured after a spin-echo period, τ–π–τ, during which either CW, TPPM or XiX proton decoupling was applied, as shown in Fig. 1. For both the CW and TPPM experiments, the acquisition and recycle delay times were 30.77 ms and 4 s respectively, while under XiX decoupling they were 25.6 ms and 5 s respectively. Note that the TPPM pulse width, τp, is generally parameterised below in terms of the corresponding nutation angle, $\theta = \tau_p \nu_1 360°$.

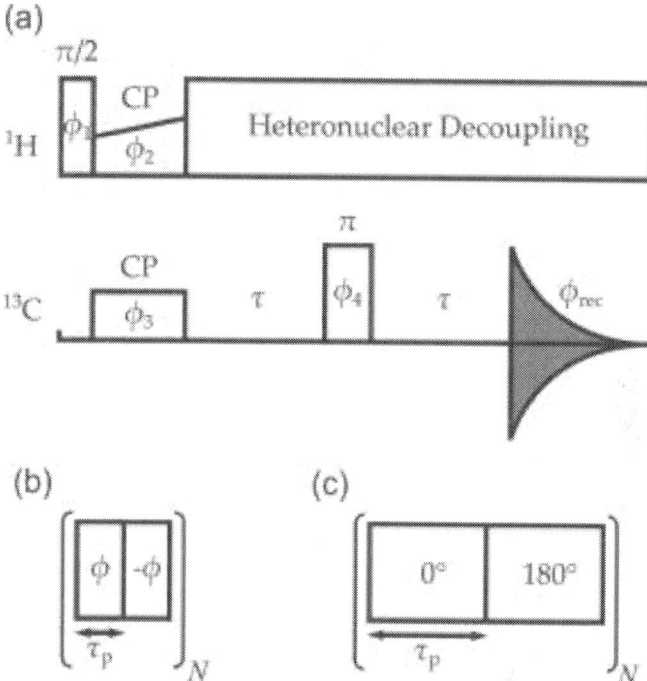

Figure 1. (a) Spin-echo pulse sequence using the same decoupling during the 2τ and acquisition periods. Phases (combining spin-temperature inversion and EXORCYCLE): $\phi_1=0°$, $180°$, $0°$, $180°$; $\phi_2=\phi_3=90°$; $\phi_4=0°$, $0°$, $90°$, $90°$; $\phi_{rec}=90°$, $270°$, $270°$, $90°$. (b) TPPM decoupling element, with phase excursion ϕ and pulse width τp. (c) XiX decoupling element.

The same ^{1}H decoupling was used in both spin-echo and acquisition periods. Although using a fixed decoupling sequence for acquisition would lead to reasonably consistent line-shapes in the acquired spectra, significant mismatches between spin-echo and acquisition decoupling were observed to distort fitted T_2 values via the orientation dependence of decoupling efficiency [38]; magnetisation that has been preserved by efficient decoupling during the spin-echo period may rapidly decohere under the acquisition decoupling, leading to an underestimate of intensity at longer spin-echo times and hence an underestimation of T_2. The variation of the orientation dependence of T_2^c with decoupling parameters is illustrated in the Supplementary Information, Fig. S2. The ^{1}H decoupling nutation rate, v_1, was measured using the same sequence with a zero-length spin-echo period, incrementing the initial ^{1}H pulse width to acquire a ^{1}H nutation spectrum and taking the peak position as the nominal v_1.

Full decay curves were obtained at selected decoupling conditions by incrementing the evolution time, 2τ , linearly in 41 steps from zero to approximately twice the maximum expected T_2'. The free induction decays were zero-filled and Fourier transformed (without apodisation) using matNMR [39]. The decay of the methylene ^{13}C peak height as a function of 2τ was fitted to a decaying exponential to obtain T_2' using MATLAB® [40]. Where detailed parameter maps as a function of the parameters of a decoupling sequence were acquired, T_2' values were inferred from a pair of experiments at $2\tau =0$ and $2\tau \approx T_{2,max}'$, assuming simple exponential decay of the peak height between these points. Discrepancies between the T_2' values obtained by this quick, but approximate, approach and those obtained from full decays were reduced by re-scaling the approximate values using a quadratic function fitted to approximate vs. accurate T_2' values at between three and five characteristic points in the parameter space. As illustrated in the SI, Fig. S1, this rescaling resulted in relatively modest changes in the T_2' values (up to 20% at maxima and 30% around minima), and allowed good T_2' estimates to be measured efficiently for a wide parameter space.

Numerical Simulations

To explore how the spin dynamics change as a function of spin-system size, spin-systems containing different numbers of protons at increasing distance from a selected methylene C atom were created, based on the room temperature neutron structure of α-glycine (CSD refcode GLYCIN20 [41]). These spin systems are labelled as CHn, with n indicating the number of protons in the system. CASTEP version 6.0 [42] was used to optimise the ^1H positions in the unit cell using a planewave cut-off energy of 600 eV. Brillouin zone integrals used a minimum sampling density of $0.1° A^{-1}$ apart with the sampling grid offset by 0.25, 0.25, and 0.25 in fractional coordinates of the reciprocal lattice. The exchange-correlation functional was approximated at the generalised-gradient level, specifically that of Perdew, Burke and Ernzerhof (PBE) [43]. Ultrasoft pseudopotentials [44] consistent with the PBE approximation were generated by CASTEP on-the-fly. Shielding tensors were subsequently calculated using the GIPAW method [45], [46] and [47].

The effects of dynamics on the dipolar and shielding tensors of protons of the NH_3^+ groups, which are in rapid exchange at ambient temperature, were accounted for by averaging the chemical shift and dipolar coupling tensors over the three H positions and diagonalising to obtain the new principal components and mean tensor orientation. Dipolar coupling tensors between the spins of the NH_3^+ were reconstructed by re-orienting the averaged dipolar tensor along the $C-NH_3^+$ bond vector and scaling by $P_2(\cos 90°) = 1/2$. This task of combining shielding tensor information from CASTEP and dipolar couplings determined from the geometry was handled with in-house software (available with the pNMRsim simulation program [48]). The dynamics of ^1H coupled networks are strongly determined by the root-sum-square of the ^1H dipolar couplings, d_{rss}, at a given site [6], and so the contributions of neglected protons outside the extracted 'cluster' of spins to d_{rss} were compensated for by scaling the ^1H homonuclear dipolar couplings so that the d_{rss} at one of the methylene ^1H sites (H5 in GLYCIN20) of the reduced spin-system matched that of the extended lattice. This d_{rss} value converges to 27.8 kHz when sufficient unit cells are considered (the value for the other methylene proton, H4, is very similar, 27.3 kHz). Note that d_{rss} for H5 without motional averaging is 30.2 kHz. The heteronuclear dipolar couplings were not scaled since the heteronuclear couplings between C_α and non-methylene protons have a negligible effect on the heteronuclear d_{rss} values. Spin systems with unscaled ^1H homonuclear couplings were also created and used in indicated cases. The ^1H chemical shift referencing was chosen to bring the methylene protons on resonance by subtracting the calculated chemical shielding values from 26.56 ppm. The ^{13}C chemical shift and the negligible J couplings were not included in the spin systems. The resulting ^{13}C, $(^1H)n$ spin systems are given the labels CHn below.

Simulations of RF decoupling under magic-angle spinning were performed in Hilbert space with pNMRsim [48], using a minimum time-step for propagator calculation of 1 µs. The theoretical background to such simulations has been extensively described elsewhere [49], [50], [51] and [52]. The simulations started with a state of ^{13}C x magnetisation and measured the remaining x magnetisation as a function of the duration of the decoupling period to create a simulated free-induction decay (FID) or spin-echo decay. In spin-echo simulations, an ideal refocusing π-pulse [53] was applied at the mid-point of the rotation-synchronised decay time. Unless otherwise indicated, powder averages were performed over all three Euler angles describing the crystallite orientation, using 150 orientations distributed over a hemisphere generated with the ZCW algorithm [54], [55] and [56]. Where the cycle times of the RF pulse sequence and sample spinning are not too dissimilar, it is generally possible to find a common time base for both the timing of the RF pulse sequence and MAS period. For phase-modulated RF pulse sequences, this allows the evolution of the density matrix to be determined from a limited number of propagators evaluated over a single period of rotation [11], greatly reducing the simulation time, usually by an order of magnitude or more. T_2 relaxation can be safely omitted from these simulations by noting that at room temperature the relaxation of the $C\alpha$ site of glycine is in the extreme narrowing limit where $T_1 = T_{1\rho} = T_2$; relaxation time constants on the order of seconds have been observed experimentally[57], much longer than the maximum T_2 observed for this site [33]. Although the ^{1}H T_1 is somewhat shorter (about 0.5–1 s), this is also orders of magnitude longer than the time constants for decay of the ^{1}H magnetisation due to "spin diffusion". The fast dynamics of the methyl group is helpful in shortening T_1 without contributing significantly to $T_{1\rho}$. When comparing time constants for coherent decay from simulation, T_2^c, with experimental T_2 values, it is important to take into account the inhomogeneity of the RF (B_1) field in typical NMR probes. The incorporation of RF inhomogeneity into the simulations is discussed in the SI.

Calculation times for the 10-spin CH$_9$ system required on average about one hour per orientation. As an example, the results for CH$_9$ data set shown in Fig. 7 were acquired on an institutional HPC cluster with the calculations for each powder orientation run in parallel on separate processors and required about 33,000 CPU hours. In contrast, the corresponding calculation times per orientation for the CH$_6$ systems were about three seconds. This is consistent with calculation times scaling as $O(N^3)$, where N is the size of the Hilbert space.

SIMULATION RESULTS

Fig. 2 shows simulated powder-averaged directly acquired and spin-echo decays under TPPM and CW decoupling, where the decoupling parameters have been chosen to result in similar rates of decay. It can be seen in Fig. 2(a) that physically reasonable decoherence is observed only when ^{1}H CSA parameters are included. This contrasts to the observation that the decay of ^{1}H magnetisation under simple MAS is essentially independent of the CSA parameters [7], but is consistent with the behaviour being largely determined by the second-order cross-terms between the heteronuclear dipolar couplings and ^{1}H CSA tensors [58], [59] and [13]. The dynamics beyond the first few milliseconds depend on the size of the spin system; both the monoexponentiality of the decays and the observed T_2^{C} tend to increase and converge to size-independent limits as the number of spins increases. The increase in T_2^{C} with increasing spin-system size is consistent with increased "self-decoupling" in a larger dipolar-coupled network [60]. These trends are similar for the two sequences.

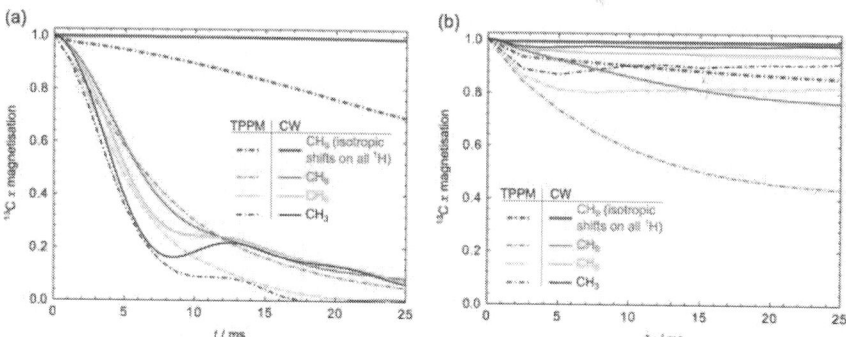

Figure 2. Simulated   powder-averaged   magnetisation   decays at $\nu_r = 12$ kHz and $\nu_0^{H} = 600$ MHz: (a) without spin echo and (b) with spin echo. Dash-dot lines correspond to TPPM decoupling at θ=198.5°, ϕ =17°, $\nu_1 = 105$ kHz and solid lines to CW decoupling at $\nu_1 = 300$ kHz. The spin systems are CH_3 (black), CH_6(cyan), CH_9 (magenta) and CH_9 using only isotropic components of ^{1}H shifts (brown). (For interpretation of the references to colour in this figure caption, the reader is referred to the web version of this paper.)

Fig. 2(b) shows the decay of ^{13}C magnetisation as a function of the spin-echo time 2τ, but with otherwise identical simulation conditions. The behaviour observed is markedly different, with incomplete decoherence of the starting magnetisation that tends towards a plateau rather than decaying to zero. Although this unphysical behaviour is reduced in the larger spin systems, the convergence is much slower than observed without the spin echo. Simulations with a spin-echo pulse of varying tip angle showed behaviour which evolved smoothly between the limits of no spin-echo and a full 180° refocusing pulse, providing some reassurance that the effects observed are not artifacts of an over-idealised simulation. The possible origins of this unexpected behaviour, which was also observed in test simulations of static samples, are discussed below.

Analogous decay curves under XiX decoupling are shown in Fig. 3. As for CW and TPPM, the T_2^C tends to increase with increasing spin-system size, with the larger spin-systems showing clearly monoexponential decays. In this case, neglecting the CSA of ^1H shift tensors (dashed lines) has a much smaller effect, which is consistent with cross-terms between the heteronuclear and the homonuclear dipolar couplings being the limiting factor for XiX decoupling away from resonance conditions [61]. The spin-echo decays under XiX, Fig. 3(b), show similar unphysical incomplete decoherence, which is gradually reduced as more spins are included in the simulation, as previously observed for CW and TPPM.

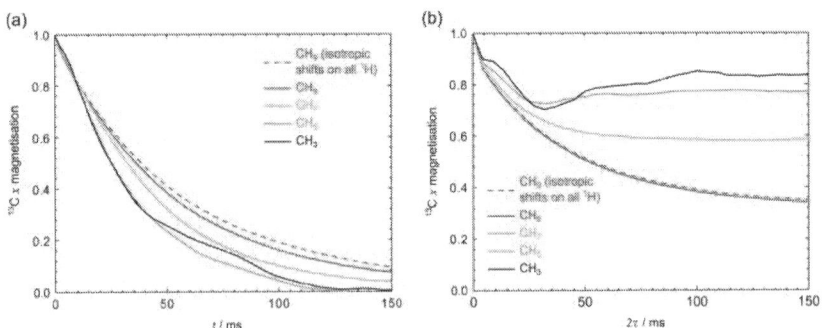

Figure 3. Simulated powder-averaged magnetisation decays under XiX decoupling at $\tau_p/\tau_r = 2.85$, $\nu_1 = 170\ kHz$, $\nu_r = 25\ kHz$ and $\nu_0^H = 500\ MHz$: (a) without spin echo and (b) with spin echo. The spin systems are CH_3(black), CH_5 (green), CH_7 (cyan), CH_9 (magenta) and CH_9 using only isotropic components of ^1H shifts (dashed magenta). (For interpretation of the references to colour in this figure caption, the reader is referred to the web version of this paper.)

The failure of the simulated spin-echo curves to decay to zero with increasing echo time is investigated in Fig. 4 using decay curves for CH_4 and CH_7 spin systems under CW decoupling for three non-special crystallite orientations. The first half, up to the π pulse at $\tau = 25$ ms, shows the ^{13}C magnetisation decaying towards zero as expected, although there is still significant oscillatory behaviour in the curves for individual orientations. These oscillations, associated with the finite size of the spin systems, are a function of the crystallite orientation, and so are effectively disguised by powder averaging, i.e. the success of the small spin-system simulations in reproducing realistic magnetisation decays should not be overplayed. As can be seen from the individual curves, the π pulse has the effect of largely reversing the oscillatory evolution, and the final amplitude, which corresponds to the $2\tau = 50$ ms point in the spin-echo simulations shown in Fig. 2(b), is almost completely refocused.

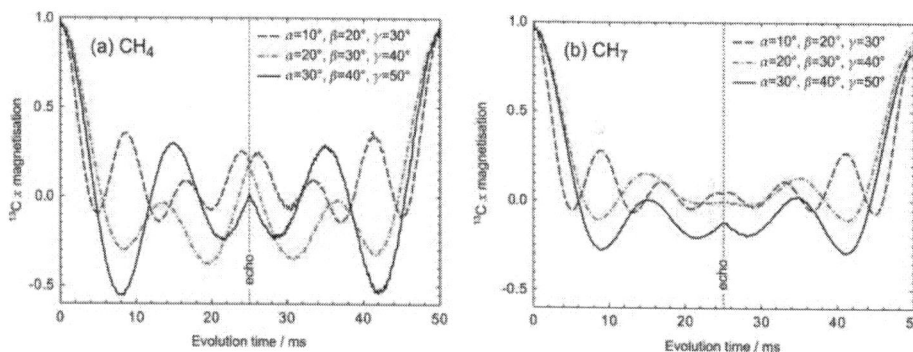

Figure 4. Simulated magnetisation decays during a spin echo for three crystallite orientations of (a) CH_4 and (b) CH_7. The dashed line in the middle of the evolution marks the position of the π -pulse. Conditions: CW decoupling at $\nu_r = 12$ kHz, $\nu_1 = 105$ kHz and $\nu_0^H = 600$ MHz. Euler angles defined as Ref. [3].

Further insight is provided by analysing the evolution of the density matrix during the spin-echo simulations. Fig. 5 shows the distribution of the matrix norm between the different ^1H spin orders as a function of time for a sample crystallite orientation under CW decoupling. The ^1H spin order for a given element of the density matrix, $\langle i|\sigma|j \rangle$, corresponds to the number of ^{1}H spins that need to be flipped to convert the bra i to the match the ket j.

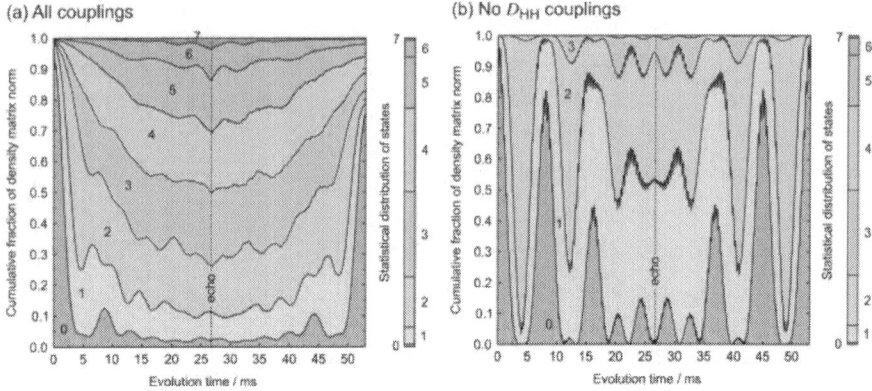

Figure 5. Square norm of the density matrix elements for each ^{1}H spin order of a single crystallite orientation (α=10°,β=20°, γ=30°) of CH_7 as a function of evolution time for CW decoupling at $\nu_1 = 105$ kHz, $\nu_r = 12$ kHz and $\nu_0^H = 600$ MHz. The black dashed line marks the position of the π-pulse. Two CH_7 spin systems are shown: (a) including all couplings and (b) without homonuclear dipolar couplings. The colour bars at the side show the spin order norms for an even distribution of magnetisation amongst all possible states. Euler angles defined as Ref. [3]. (For interpretation of the references to colour in this figure caption, the reader is referred to the web version of this paper.)

Fig. 5(a) shows that the populated states initially all have a ^{1}H spin order of zero (corresponding to the starting state of pure ^{13}C magnetisation), but the higher spin orders are quickly populated due to "spin-diffusion" driven by the homonuclear ^{1}H couplings. The distribution of magnetisation immediately before the π pulse is close to statistical (i.e. reflecting the relative numbers of states of given order), but this evolution is mostly reversed by the π pulse. This behaviour is much more marked in the artificial system, Fig. 5(b), in which homonuclear couplings have been removed. Although the initial decay of spin order 0, due to the cross-term between the heteronuclear dipolar coupling and ^{1}H CSA, is the same in (a) and (b), the subsequent evolution shows much stronger oscillations, with the higher spin orders being populated much more slowly, and the starting ^{13}C magnetisation is fully refocussed. It is clear that the homonuclear couplings have a significant impact on the dynamics, but the underlying behaviour is still driven by the dipolar/CSA cross-term.

The unphysical refocusing of the magnetisation can only be an artifact of the finite size of the spin system, and has a significant impact on the utility of simulations based on finite spin-systems. Such simulations can only be guaranteed to reflect experimental observations at short timescales before the higher spin orders are populated and the "phase space" of the simulations has been filled. In the real spin-system, the density matrix norm can spread indefinitely into an infinite phase space, and a refocusing π pulse is unable to reverse this evolution. The essentially coherent nature of the evolution is less obvious in the absence of a refocusing pulse and the small system simulations are surprisingly effective [7], particularly when the evolution is averaged over multiple orientations.

The role of high spin orders in the spin dynamics raises the question of the extent to which decoupling performance is influenced by the parameters of distant spins. Fig. 6shows simulated T_2^c values as a function of the TPPM pulse width using three model CH_8spin systems: one (solid line) the normal glycine-derived CH_8 system, the second (dash-dot line) with only isotropic chemical shifts on distant protons (further than the two methylene ^1H spins bonded to the ^{13}C), and the third (dashed line) without heteronuclear dipolar couplings to those distant protons. When T_2^c is small, i.e. the magnetisation decays quickly, the parameters of the remote spins have negligible impact. Around regions of peak decoupling, however, very different decay rates are observed. Unsurprisingly, given the mode of action of TPPM, including the CSA and heteronuclear dipolar couplings to the remote spins (solid line) results in poorer performance. Very similar effects of distant-spin parameters are observed across XiX parameter maps, except these are more sensitive to distant heteronuclear couplings rather than ^1H CSAs, see Fig. S3 in the SI. Although these distant heteronuclear couplings contribute negligibly to the total d_{rss}, their impact on XiX decoherence times is significant. This makes it difficult to make quantitative predictions of peak decoupling since the dynamics clearly depend on both a large number of spin parameters and having a large numbers of spins i.e. it is not sufficient to 'rescale' the parameters of a small spin-system simulation.

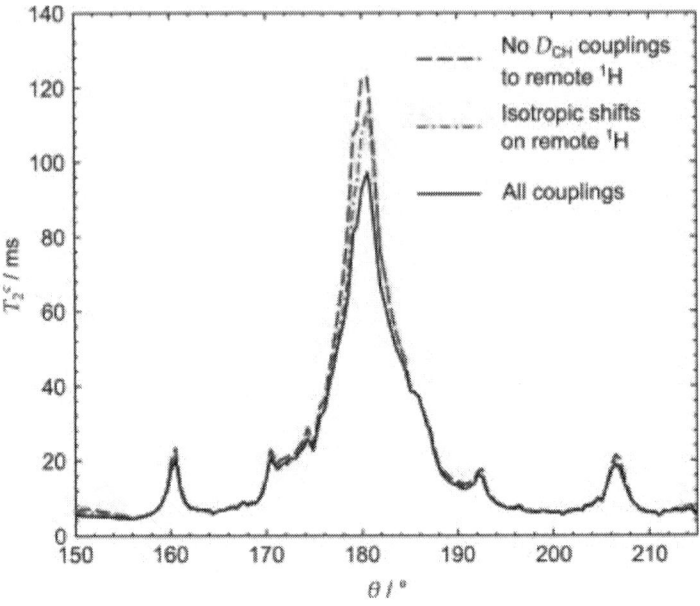

Figure 6. Simulated TPPM parameter cross-sections as a function of TPPM pulse tip-angle, θ, at $\phi = 7°$, $\nu_r = 25\,\text{kHz}$, $\nu_1 = 170\,\text{kHz}$ and $\nu_0^H = 500\,\text{MHz}$. Three variations of the CH_8 spin system are shown: (dashed) heteronuclear dipolar couplings to only the two most-central methylene protons, (dash-dot) isotropic chemical shifts on all but the two most-central methylene protons, and (solid) all couplings included.

CAN DECOUPLING PERFORMANCE BE PREDICTED?

These results demonstrate that decay of nuclear spin magnetisation under heteronuclear decoupling can only be effectively simulated using small spin-systems in a relatively narrow set of circumstances, for example, simulations of magnetisation decays under poor TPPM decoupling, Fig. 6, while simulating T_2 decays is even more challenging due to unphysical refocusing effects. As discussed in the introduction, state-space restriction methods [17], [18], [21], [20] and [25] have recently been developed that allow much larger numbers of spins to be simulated, at the expense of neglecting higher spin orders. This is effective in solution-state NMR where the spin–spin couplings are relatively weak and high-order coherences relax quickly [22] — the high order coherences do not build up sufficient population to have much impact on the spin dynamics. However, in general, this approach is not suitable for solids due to the strong couplings and slower relaxation. For glycine in the solid state we can infer that $T_2 = T_1 = 5$ s, as discussed previously. Given that 12-spins are about the limit for Hilbert-space simulation, we can estimate the maximum spin–spin couplings that would allow accurate simulation of the spin dynamics for this relaxation rate. Following the procedure of Ref. [22], the couplings would have to be no more than 2.1 Hz if the relaxation rates are proportional to the coherence order, n, as in solution-state. This value is at least an order of magnitude smaller than the effective couplings observed in typical organic solids under MAS, as shown by the time constants for decay of the ^{13}C magnetisation of 1–2 ms observed in Fig. 5. We assume that the scaling of the relaxation rate with coherence order is the same in molecular solids and liquids based on experimental observations that relaxation in glycine is in the fast-motion limit [57], and noting that the dynamics in the solid state are not dissimilar to those in the solution state but just without translational degrees of freedom that are irrelevant to relaxation (with plastic crystals such as adamantane being an extreme example). If, however, we assume as Karabanov et al. [22] that the relaxation rates are proportional to $\sqrt{n}$, then even smaller spin–spin couplings of less than 0.7 Hz would be needed for a realistic 12-spin simulation. See the SI for further discussion of relaxation in solids.

Nevertheless, restricted state-space simulations have had some success for solid-state simulations in the absence of RF decoupling [23] and [24]. Similarly "effective field" approaches have been used [60] to describe the effects of decay of coherence into the ^{1}H bath, but at the expense of introducing additional empirical parameters. We can somewhat crudely mimic these Liouville space simulations within the confines of a classical Hilbert space calculation by repeatedly nulling the amplitudes of higher spin orders in the density matrix, e.g. every rotor period. This might approximate the behaviour of an extended spin-system, with the magnetisation quickly evolving to higher spin orders and never returning.

However, it was found (results not shown) that periodically nulling just the highest spin order had little effect on the unphysical refocusing. Nulling more spin orders progressively reduced the amount of refocusing, but, by effectively introducing relaxation to the decoherence, this artificially shortened the decay and so failed to reproduce the long-term evolution.

It is evident from previous literature [32], [59] and [13] and the results above that simulations involving relatively few spins, 2–4, can reproduce the qualitative performance of decoupling sequences with respect to their parameters. However, the quantitative values of decay rates obtained from simulation depend strongly on the size of the spin system and the various parameters involved. Fig. 7(a) shows a cross-section of the TPPM parameter space at a fixed phase excursion as a function of spin system size. Larger than about CH_4, the patterns are more-or-less consistent, which suggests it might be possible to "rescale" the results from calculations on a small system onto those obtained from much costlier multi-spin calculations. This is illustrated in Fig. 7(b) and (c), which plot the fitted T_2^C values from simulations of one spin-system against corresponding ones obtained using a different spin-system. Although the decoherence times observed in CH_2 bear little correlation to those in CH_9, there is a close-to-linear relationship between the results obtained from CH_6 and CH_9 systems. Given that each additional spin reduces the calculation efficiency by close to an order of magnitude, there seems little value in performing a calculation on a CH_9 system if very similar results can be obtained by rescaling (by a factor of $\gamma=1.58$) results obtained on simulations of a CH_6 system. In principle, more complex analytical "transfer functions" could be used to perform this rescaling, but simple linear functions are adequate here.

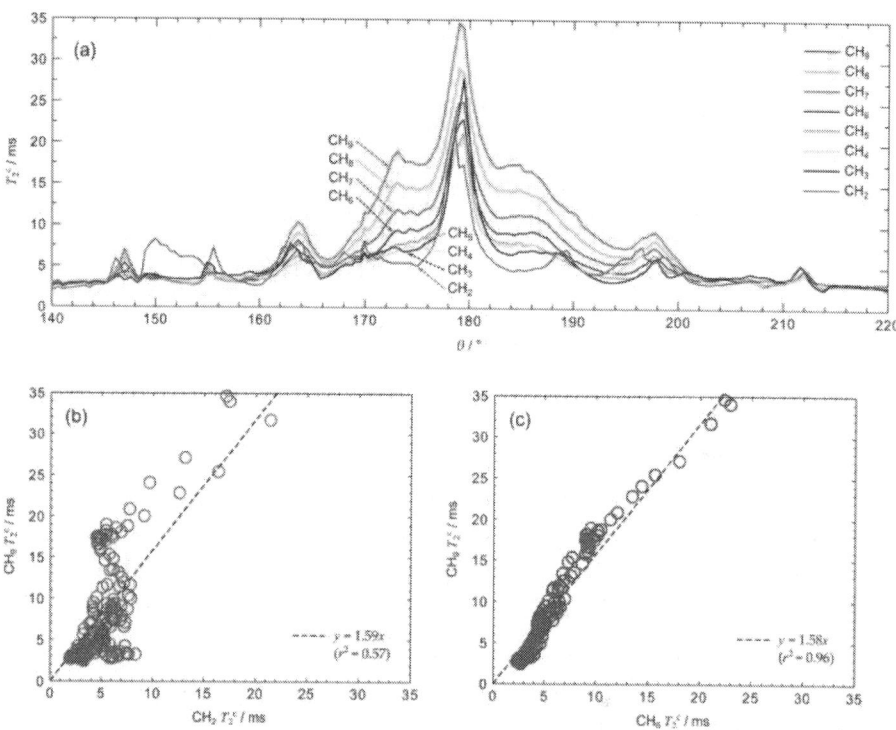

Figure 7. (a) Simulated TPPM parameter cross-sections as a function of spin-system size and TPPM pulse tip-angle, θ, at ϕ $=6°$, $\nu_r = 12$ kHz, $\nu_1 = 105$ kHz and $\nu_0^H = 600$ MHz. (b, c) An example of two transfer functions between two pairs of spin systems, with the dashed lines representing fits to $y=\gamma x$. Each circle corresponds to one of the 201 data points in (a). The correlation coefficients squared, r^2, for the fits are indicated in parentheses.

Fig. 8(a) show the results of plotting the correlations illustrated in Fig. 7(b) and (c), and extracting a transfer function gradient, γ, for different pairs of spin systems. Trivially, the gradient tends towards unity as the difference in the number of spins between the two systems decreases to zero. Simulations using fewer than six spins correlate poorly with larger simulations. On the other hand, when more distant ^{1}H spins are added, the transfer function gradients and correlation coefficients both steadily tend towards unity. The CH_5 system corresponds to the methylene carbon plus all the hydrogen atoms of a single glycine molecule, but it is not safe to assume that this explains the strong vs. erratic degree of correlation for spin systems that are larger vs. smaller than five ^{1}H spins. The number of spins necessary to start seeing simulation results representative of bulk behaviour is expected to depend on the decoupling performance in that region of the parameter space, with smaller numbers of spins needed to reproduce the T_2^C decay in regions of poor decoupling. The dashed line in Fig. 8(a) shows the mapping onto the CH$_9$results where the homonuclear d_{rss} has not been scaled to match the limit of an infinitely large system. Although the scaling factors are slightly different, the trends are the same, and so there is little advantage, in this case, of scaling the strength of the homonuclear couplings to match the extended lattice. It is worth noting that the transfer gradient increases as the homonuclear coupling network is strengthened in going from the unscaled to the scaled d_{rss}, whereas it decreases when the size of the spin system increases. This suggests that it is the increased size of the Hilbert space, rather than the couplings themselves, that are responsible for the trends with increasing spin system size. The comparison of transfer function gradients for XiX, Fig. 8(b), shows the same overall trends as for TPPM, although there is a noticeable even–odd alternation of the gradient values, which is slowly damped as the size of the space increases. This has some similarities with the observation by Halse et al. of very strong even–odd alternation when varying the maximum allowed spin order in simulations of ^{1}H spin diffusion in a static solid [25].

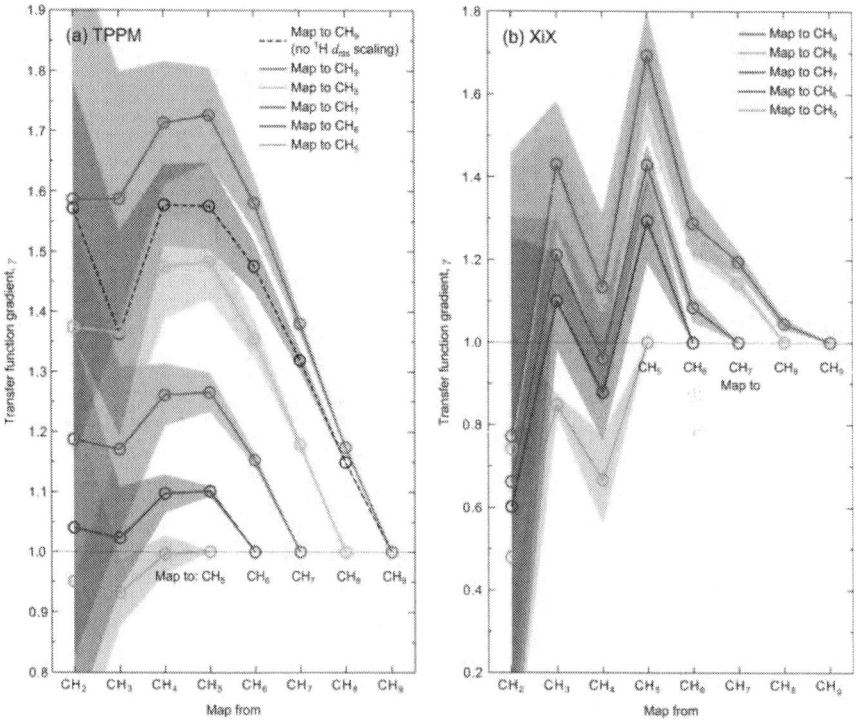

Figure 8. Transfer function gradients for pairs of spin systems for (a) TPPM decoupling (conditions as Fig. 7) and (b) XiX decoupling ($\nu_r = 25$ kHz, $\nu_1 = 170$ kHz and $\nu_0^H = 500$ MHz). The shaded regions indicate the linearity of the transfer functions, as measured by r^2, as a pseudo-error bar, $\pm \epsilon$ where $\epsilon = 1 - r^2$. The dashed line in (a) corresponds to spin systems where the homonuclear dipolar couplings were *not* scaled to match the d_{rss} of the extended structure.

Fig. 9(a) compares fitted T_2^C values obtained from the five largest spin-system simulations of Fig. 7(a), and the experimental T_2 recorded under the same conditions, as described in the Experimental section. The good agreement between the shape of the simulated and experimental parameter maps is reflected in Fig. 9(c), which shows the mapping between the simulated T_2^C decay constants to corresponding experimental T_2 values. Note how the very strong linear correlation between experimental T_2 and simulated T_2^C weakens as T_2^C increases beyond about 8 ms. Comparison with Fig. 5(a) suggests that this corresponds to the point where the finite size of the spin system has an increasing impact. The fact that the slope at smaller T_2^c is close to unity, however, is of limited significance; other slices through the parameter surfaces show the same trends, but with different slopes observed.

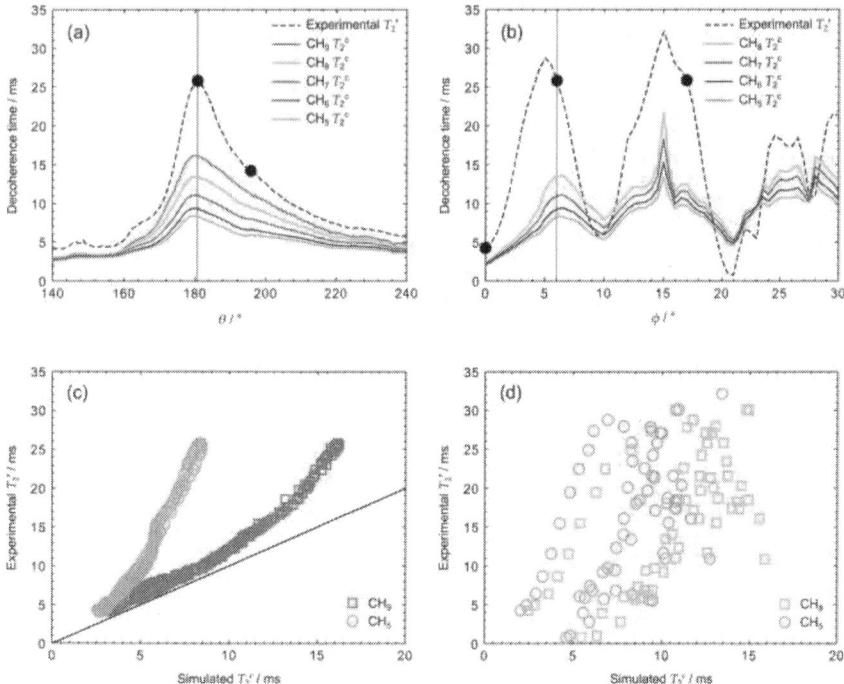

Figure 9. Cross-sections through the TPPM parameter map as a function of spin system size and TPPM pulse tip-angle, θ, at $\nu_r = 12$ kHz, $\nu_1 = 105$ kHz and $\nu_0^H = 600$ MHz: (a) $\phi = 6°$ and a range of pulse widths, (b)$\theta = 178.5°/\cos(\phi/1.2)$ and a range of pulse phases. The vertical grey line marks the intersection of the two cross-sections. Approximate T_2' values inferred from parameter map points at $2\tau = 0, 15$ ms and $2\tau = 0, 20$ ms for (a) and (b) respectively. The impact of ν_1 inhomogeneity has been taken into account in the simulated results using a quick, approximate method and more exactly in (a) and (b) respectively. Full T_2' decays were measured at the points marked with • and the parameter map scaled to fit. Transfer functions mapping some of the simulations onto the experimental results are shown in (c) and (d) for the results in (a) and (b) respectively. The straight line in (c) marks $T_2^c = T_2'$ as a guide to the eye.

The comparison as a function of TPPM pulse phase, along a line passing through several local optima close to the theoretical optimum $\theta = 180°/\cos(\phi)$[13], Fig. 9(b) and (d), is much poorer. Both sets of transfer functions in Fig. 9(c) and (d) show the same pattern of convergence towards the experimental results, with largest deviations being observed for the longest decoherence times. The much weaker correlation observed in Fig. 9(d) is associated with the significantly greater complexity of this cross-section through the parameter surface; it includes regions of very poor decoupling, which are not well characterised by the choice of sampling point for the T_2 decays, combined with regions of good performance which require large spin-systems to be accurately described. Note that the corrections for RF inhomogeneity have little impact on the depth of the minima around ϕ =10° and 20°, and so the discrepancies in this region are more likely to be associated with inaccurate measurements of very short T_2 values rather than deficiencies of the calculations. Similarly poor correlations when multiple resonance conditions are involved are also observed for XiX decoupling (Fig. S5 in the Supplementary Information).

It is worth considering whether the disagreement in Fig. 9(d) is related to the differences between the experiments and the simulations. Firstly the experiments measure T_2, using a spin-echo to refocus inhomogeneous contributions to the decay rate (which are not present in simulation), but are compared to simulated T_2^c values, since T_2 simulations are complicated by unphysical behaviour, cf. Fig. 2. All the experimental evidence suggests that the positions of T_2^c and T_2 optima are the same [62]. This was confirmed in correlation plots using experimental peak height as the metric of experimental performance (Figs. S6 and S7 in the Supplementary Information), which show the same behaviour observed in Fig. 9. The simulations also assume that the initial ^{13}C magnetisation is uniform across all crystallite orientations. Simulations of the orientation-dependence of the CP efficiency, however, show that ^{1}H spin-diffusion during the contact times used (1.2 ms and 2.7 ms) results in close to uniform excitation. The final point of difference is that the simulations sample the total remaining ^{13}C magnetisation at the end of the decay period, but before acquisition, while the experimental measurement is based on the peak height of the methylene signal. As discussed in the Experimental section, using the same decoupling during spin-echo and acquisition periods minimises cases where magnetisation that has been retained during the decay period rapidly dephases and is not observed in acquisition period.

CONCLUDING REMARKS

We have tested the fundamental limitations of exact simulations of the heteronuclear decoherence times T_2^c (directly observed FID) and T_2' (under a spin-echo) in the solid state under simultaneous MAS and RF decoupling. Counterintuitively, a spin-echo pulse on the observed nucleus is found to largely refocus the decay of ^{13}C magnetisation. Increasing the size of the spin system reduces the degree of refocusing, but it is not possible to reproduce experimental T_2' decays using the largest number of coupled spins that can be simulated exactly. Attempts to eliminate this unphysical refocusing by suppressing higher order coherences (mimicking simulations in which the state space is restricted to lower spin orders) were unsuccessful. Turning this around, however, we predict that liquid-crystalline materials, where the spin systems are limited to the liquid crystal molecules, would show this refocusing effect in spin-echo experiments.

In contrast, physically realistic T_2^c decays are obtained in simulation, but the nuclear spin decoherence around regions of good decoupling is found to be very sensitive to the size of the spin system and the details of the various spin-system parameters, including ^{1}H CSA parameters and/or dipolar couplings of distant spins. This makes quantitative prediction of peak decoupling performance particularly problematic. High spin-order coherences are rapidly populated, and while the experimental behaviour can be qualitatively reproduced by mimicking the effects of ^{1}H spin diffusion [1], it seems unlikely that simulations within a small Hilbert or Liouville space can quantitatively reproduce experimental behaviour without invoking adjustable and purely empirical parameters. This contrasts to earlier work that found that modest numbers (9–10) of ^{1}H spins were sufficient to predict the evolution of ^{1}H magnetisation under MAS [7] without the need for adjustable parameters. The difference is presumably related to the shorter timescales for the decoherence of the ^{1}H spin order compared to the much slower decay of ^{13}C magnetisation under RF decoupling.

Investigating the dependence of T_2^c decays on spin-system size, we find that the results for simulations in systems of 6–7 spins can often be usefully mapped on to those from larger, costlier calculations. Further increasing the size of the system leads to a monotonic rescaling of the decoherence times, and allows longer decoherence times to be observed at decoupling optima, but without significantly altering the map of performance vs. decoupling parameters. This confirms that an important determining factor in decoherence times is the size of "spin space" into which the magnetisation can spread. However, these 'transfer functions' tend to vary between different parts of the parameter space and between decoupling sequences, limiting their applicability to predict quantitatively the performance of arbitrary sequences. This also implies that state-space restriction techniques will similarly struggle to reproduce the quantitative behaviour of many-spin systems under RF decoupling.

Mapping of simulation results onto experimental behaviour is significantly less successful than mapping between simulation results in differently sized spin systems. As seen in Fig. 9, simulations correlate very well with experimental behaviour in smoothly varying regions of the parameter space (subject to the limitations of a finite simulation space discussed above). In regions where the parameter space is rapidly changing, however, there is not a straightforward mapping between experimental results and simulation. Considering Fig. 6, it is reasonable to suppose that the behaviour in rapidly changing regions of the parameter space is highly dependent on both long- and short-range NMR parameters as well as instrumental factors, such as RF inhomogeneity. In the case of sequences such as XiX, the optimum decoupling conditions result from the interaction of multiple resonance conditions; quantitative prediction of decoupling performance in this situation is likely to be extremely challenging.

On the other hand, the optimum decoupling *parameters* are robustly reproduced using a modest number of spins, and the structure of the parameter maps is not sensitive to the parameters of remote spins. This is consistent with the routine experimental practice of optimising decoupling on a set-up system and using the same decoupling sequence parameters for the sample under study; it would not be possible to optimise experiments in this way if the positions of decoupling optima were strongly dependent on the parameters of multiple spins. Similarly, it is quite practical to perform multi-variable parametrisations of the performance of a decoupling sequence using a 6–7 spin system under different conditions by exploiting the efficient simulation techniques used here. While quantitative prediction of the performance of different local optima will remain a major challenge, this can be avoided in experimental practice by direct optimisation of the decoupling parameters [30].

ACKNOWLEDGMENTS

This work was funded by Engineering and Physical Sciences Research Council grant EP/H023291/1, and benefitted from discussions within the context of the EPSRC-funded collaborative computational project for NMR crystallography (CCPNC).

REFERENCES

1. M. Ernst, J. Magn. Reson. 162 (1) (2003) 1–34. http://dx.doi.org/10.1016/S1090- 7807(03)00074-0.
2. P. Hodgkinson, Prog. Nucl. Magn. Reson. Spectrosc. 46 (2005) 197–222. http: //dx.doi.org/10.1016/j.pnmrs.2005.04.002.
3. P. Hodgkinson, L. Emsley, Prog. Nucl. Magn. Reson. Spectrosc. 36 (2000) 201–239. http://dx.doi.org/10.1016/S0079-6565 (99)00019-9.
4. M. Bak, J.T. Rasmussen, N.C. Nielsen, J. Magn. Reson. 147 (2) (2000) 296–330. http://dx.doi.org/10.1006/jmre.2000.2179.
5. M. Veshtort, R.G. Griffin, J. Magn. Reson. 178 (2) (2006) 248–282. http://dx.doi. org/10.1016/j.jmr.2005.07.018.
6. V.E. Zorin, S.P. Brown, P. Hodgkinson, Mol. Phys. 104 (2) (2006) 293–304. http: //dx.doi.org/10.1080/00268970500351052.
7. V.E. Zorin, S.P. Brown, P. Hodgkinson, J. Chem. Phys. 125 (2006) 144508.
8. G. de Paëpe, D. Sakellariou, P. Hodgkinson, S. Hediger, L. Emsley, Chem. Phys. Lett. 368 (2003) 511–522.
9. M. Leskes, R.S. Thakur, P.K. Madhu, N.D. Kurur, S. Vega, J. Chem. Phys. 127 (2) (2007) 024501. http://dx.doi.org/10.1063/1.2746039.
10. J.M. Griffin, C. Tripon, A. Samoson, C. Filip, S.P. Brown, Magn. Reson. Chem. 45 (S1) (2007) S198–S208. http://dx.doi.org/10.1002/mrc.2145.
11. V.E. Zorin, M. Ernst, S.P. Brown, P. Hodgkinson, J. Magn. Reson. 192 (2) (2008) 183–196. http://dx.doi.org/10.1016/j.jmr.2008.02.012.

12. R.S. Thakur, N.D. Kurur, P.K. Madhu, J. Magn. Reson. 193 (1) (2008) 77–88. http://dx.doi.org/10.1016/j.jmr.2008.04.024.

13. I. Scholz, P. Hodgkinson, B.H. Meier, M. Ernst, J. Chem. Phys. 130 (2009) 114510.

14. A.S. Tatton, I. Frantsuzov, S.P. Brown, P. Hodgkinson, J. Chem. Phys. 136 (2012) 084503. http://dx.doi.org/10.1063/1.3684879.

15. V.S. Mithu, P.K. Madhu, Chem. Phys. Lett. 556 (2013) 325–329. http://dx.doi. org/10.1016/j.cplett.2012.11.016.

16. P. Hodgkinson, D. Sakellariou, L. Emsley, Chem. Phys. Lett. 326 (5–6) (2000) 515–522. http://dx.doi.org/10.1016/S0009-2614 (00)00801-0.

17. I. Kuprov, N. Wagner-Rundell, P.J. Hore, J. Magn. Reson. 189 (2) (2007) 241–250. http://dx.doi.org/10.1016/j.jmr.2007.09.014.

18. I. Kuprov, J. Magn. Reson. 195 (1) (2008) 45–51. http://dx.doi.org/10.1016/j. jmr.2008.08.008.

19. M.C. Butler, J.-N. Dumez, L. Emsley, Chem. Phys. Lett. 477 (2009) 377–381.

20. J.-N. Dumez, M.C. Butler, L. Emsley, J. Chem. Phys. 133 (2010) 224501. http: //dx.doi.org/10.1063/1.3505455.

21. H.J. Hogben, P.J. Hore, I. Kuprov, J. Chem. Phys. 132 (17) (2010) 174101. http: //dx.doi.org/10.1063/1.3398146.

22. A. Karabanov, I. Kuprov, G.T.P. Charnock, A. van der Drift, L.J. Edwards, W. Kockenberger, J. Chem. Phys. 135 (8) (2011) 084106. http://dx.doi.org/ 10.1063/1.3624564.

23. J.-N. Dumez, M.C. Butler, E. Salager, B. Elena-Herrmann, L. Emsley, Phys. Chem. Chem. Phys. 12 (2010) 9172–9175. http://dx.doi.org/10.1039/ C0CP00050G.

24. J.-N. Dumez, M.E. Halse, M.C. Butler, L. Emsley, Phys. Chem. Chem. Phys. 14 (2012) 86–89. http://dx.doi.org/10.1039/C1CP22662B.

25. M.E. Halse, J.-N. Dumez, L. Emsley, J. Chem. Phys. 136 (2012) 224511. http://dx. doi.org/10.1063/1.4726162.

26. H. Geen, R. Graf, A.S. Heindrichs, B.S. Hickman, I. Schnell, H.W. Spiess, J.J. Titman, J. Magn. Reson. 138 (1) (1999) 167–172. http://dx.doi.org/ 10.1006/jmre.1999.1711.

27. C.E. Hughes, Prog. Nucl. Magn. Reson. Spectrosc. 45 (3–4) (2004) 301–313. http://dx.doi.org/10.1016/j.pnmrs.2004.08.002.

28. H.G. Krojanski, D. Suter, Phys. Rev. A 74 (2006) 062319. http://dx.doi.org/ 10.1103/PhysRevA.74.062319.

29. A. Lesage, C. Auger, S. Caldarelli, L. Emsley, J. Am. Chem. Soc. 119 (33) (1997) 7867–7868.

30. G. de Paëpe, N. Giraud, A. Lesage, P. Hodgkinson, A. Böckmann, L. Emsley J. Am. Chem. Soc. 125 (46) (2003) 13938–13939. http://dx.doi.org/10.1021/ja037213j PMID: 14611212.

31. D.C. Apperley, R.K. Harris, P. Hodgkinson, Solid-State NMR: Basic Principles & Practice, Momentum Press, New York (2012) 109–140 (Chapter 5).

32. A. Detken, E.H. Hardy, M. Ernst, B.H. Meier, Chem. Phys. Lett. 356 (3–4) (2002) 298–304. http://dx.doi.org/10.1016/S0009-2614 (02)00335-4.

33. S.K. Vasa, H. Janssen, E.R.H. Van Eck, A.P.M. Kentgens, Phys. Chem. Chem. Phys. 13 (2010) 104–106. http://dx.doi.org/10.1039/C0CP01929A.

34. R.E. Taylor, Concepts Magn. Reson. A 22 (2) (2004) 79–89. http://dx.doi.org/ 10.1002/cmr.a.20015.

35. C.E. Hughes, K.D.M. Harris, Chem. Commun. 46 (2010) 4982–4984. http://dx. doi.org/10.1039/C0CC01007C.

36. E.V. Boldyreva, V.A. Drebushchak, T.N. Drebushchak, I.E. Paukov, Y.A. Kovalevskaya, E.S. Shutova, J. Therm. Anal. Calorim. 73 (2) (2003) 419–428. http://dx.doi.org/10.1023/A:1025457524874.

37. A.E. Bennett, C.M. Rienstra, M. Auger, K.V. Lakshmi, R.G. Griffin, J. Chem. Phys. 103 (16) (1995) 6951–6958. http://dx.doi.org/10.1063/1.470372.

38. V.E. Zorin, B. Elena, A. Lesage, L. Emsley, P. Hodgkinson, Magn. Reson. Chem. 45 (S1) (2007) S93–S100. http://dx.doi.org/10.1002/mrc.2108.

39. J.D. van Beek, J. Magn. Reson. 187 (1) (2007) 19–26. http://dx.doi.org/10.1016/j. jmr.2007.03.017.

40. MATLAB Release R2013b, The MathWorks Inc., Natick, MA, USA.

41. P. Langan, S.A. Mason, D. Myles, B.P. Schoenborn, Acta Crystallogr. B 58 (4) (2002) 728–733. http://dx.doi.org/10.1107/S0108768102004263.

42. S.J. Clark, M.D. Segall, C.J. Pickard, P.J. Hasnip, M.I.J. Probert, K. Refson, M.C. Payne, Z. Kristallogr. 220 (2005) 567–570. http://dx.doi.org/10.1524/zkri.220.5.567.65075.

43. J.P. Perdew, K. Burke, M. Ernzerhof, Phys. Rev. Lett. 77 (1996) 3865–3868. http: //dx.doi.org/10.1103/PhysRevLett.77.3865.

44. D. Vanderbilt, Phys. Rev. B 41 (1990) 7892–7895. http://dx.doi.org/10.1103/PhysRevB.41.7892.

45. C.J. Pickard, F. Mauri, Phys. Rev. B 63 (2001) 245101. http://dx.doi.org/10.1103/ PhysRevB.63.245101.

46. R.K. Harris, P. Hodgkinson, C.J. Pickard, J.R. Yates, V. Zorin, Magn. Reson. Chem. 45 (S1) (2007) S174–S186. http://dx.doi.org/10.1002/mrc.2132.

47. C. Bonhomme, C. Gervais, F. Babonneau, C. Coelho, F. Pourpoint, T. Azaïs, S.E. Ashbrook, J.M. Griffin, J.R. Yates, F. Mauri, C.J. Pickard, Chem. Rev. 112 (11) (2012) 5733–5779. http://dx.doi.org/10.1021/cr300108a.

48. P. Hodgkinson, pNMRsim: A General Simulation Program for Large Problems in Solid-State NMR, URL: ⟨http://www.durham.ac.uk/paul.hodgkinson/ pNMRsim⟩.
49. P. Hodgkinson, L. Emsley, Prog. Nucl. Magn. Reson. Spectrosc 36 (3) (2000) 201–239.
50. M. Bak, J.T. Rasmussen, N.C. Nielsen, J. Magn. Reson. 147 (2000) 296–330.
51. M. Edén, Concepts Magn. Reson. 17A (2003) 117–154.
52. M. Veshtort, R.G. Griffin, J. Magn. Reson. 178 (2006) 248–282.
53. N.S. Barrow, J.R. Yates, S.A. Feller, D. Holland, S.E. Ashbrook, P. Hodgkinson, S.P. Brown, Phys. Chem. Chem. Phys. 13 (2011) 5778–5789. http://dx.doi.org/ 10.1039/C0CP02343D.
54. S.K. Zaremba, Ann. Mat. Pura Appl. 73 (1966) 293–317. http://dx.doi.org/ 10.1007/BF02415091.
55. H. Conroy, J. Chem. Phys. 47 (12) (1967) 5307–5318. http://dx.doi.org/ 10.1063/ 1.1701795.
56. V.B. Cheng, H.H. Suzukawa Jr., M. Wolfsberg, J. Chem. Phys. 59 (8) (1973) 3992–3999. http://dx.doi.org/10.1063/1.1680590.
57. A. Krushelnitsky, R. Kurbanov, D. Reichert, G. Hempel, H. Schneider, V. Fedotov, Solid State Nucl. Magn. Reson. 22 (4) (2002) 423–438. http://dx.doi.org/10.1006/snmr.2002.0071.
58. M. Ernst, S. Bush, A.C. Kolbert, A. Pines, J. Chem. Phys. 105 (9) (1996) 3387–3397. http://dx.doi.org/10.1063/1.472224.
59. M. Ernst, A. Samoson, B.H. Meier, J. Chem. Phys. 123 (6) (2005) 064102. http: //dx.doi.org/10.1063/1.1944291.
60. M. Ernst, A. Verhoeven, B.H. Meier, J. Magn. Reson. 130 (1998) 176–185.
61. M. Ernst, H. Geen, B.H. Meier, Solid State Nucl. Magn. Reson. 29 (1–3) (2006) 2–21. http://dx.doi.org/10.1016/j.ssnmr.2005.08.004.
62. G. de Paëpe, B. Elena, L. Emsley, J. Chem. Phys. 121 (7) (2004) 3165–3180.
63. P. Tekely, P. Palmas, D. Canet, J. Magn. Reson. Ser. A 107 (1994) 129–133. http: //dx.doi.org/10.1006/jmra.1994.1059.

CITATION

Ilya Frantsuzov, Matthias Ernst, Steven P. Brown, Paul Hodgkinson, Simulating spin dynamics in organic solids under heteronuclear decoupling, Solid State Nuclear Magnetic Resonance, Available online 15 May 2015, ISSN 0926-2040, http://dx.doi.org/10.1016/j.ssnmr.2015.05.003.

CHAPTER 8

Structure and Properties of Solid-state Synthesized Poly (3,4 Propylenedioxythiophene)/Nano-ZnO Composite

Ahmat Ali[1, 2], Ruxangul Jamal[1, 2], Weiwei shao[1, 2], Adalet Rahman[1], Yakupjan Osman[1, 2] and Tursun Abdiryim[1, 2]

[1]Key Laboratory of Petroleum and Gas Fine Chemicals, Educational Ministry of China, College of Chemistry and Chemical Engineering, Xinjiang University, Urumqi 830046, China
[2]Key Laboratory of Functional Polymers, Xinjiang University, Urumqi 830046, China

ABSTRACT

Poly(3,4-propylenedioxythiophene)/nano-Zinic Oxide (PProDOT/ZnO) composites with the content of 3–7 wt% nano-ZnO were synthesized by the solid-state method with $FeCl_3$ as oxidant. The structure and morphology of the composites were characterized by Fourier transform infrared (FTIR) spectroscopy, ultraviolet–visible (UV–vis) absorption spectroscopy, X-ray diffraction (XRD) and transmission electron microscopy (TEM). The electrochemical performances of the composites were investigated by galvanostatic charge–discharge, cyclic voltammetry and electrochemical impedance spectroscopy (EIS). The photocatalytic activities of the composites were investigated by the degradation of methylene blue (MB) dyes in aqueous medium under UV light irradiation. The results from FTIR and UV–vis spectra showed that the PProDOT/ZnO composites were successfully synthesized by solid-state method, and nano-ZnO had great influences on the conjugation length and oxidation degree of the polymers. Furthermore, the PProDOT/5 wt%ZnO had the highest conjugation and oxidation degree among the composites. The results of XRD analysis indicated that there were some $FeCl_4^-$ ions as doping agent in the PProDOT matrix, and the content of ZnO had no effect on diffraction pattern of PProDOT. Morphological studies revealed that the pure PProDOT and composites had similar morphological structure, and all the composites displayed an irregular sponge like morphology. The results of electrochemical tests showed that the PProDOT/5 wt%ZnO had a higher electrochemical activity with a specific capacitance value of 220 F g^{-1} than others. The results from photocatalytic activities of the composites indicated that the PProDOT/5 wt%ZnO had better photocatalytic activity than other composites.

INTRODUCTION

Over past few decades, most of the studies of conducting polymers that include polyaniline [1], polypyrrole [2], and polythiophene [3] have been investigated [4]. 3,4-alky-lenedioxythiophenes have been studied extensively due to their good environmental stability, thermal and chemical stability, high conductivity, high transparency and low oxidation potential and used in broad applications such as batteries, super capacitors, sensors, electronic devices, and corrosion protection in organic coatings [5] and [6]. The 3,4-Propylenedioxythiophene (ProDOT), as a derivative of the 3,4-ethylenedioxy-thiophene (EDOT), shares attractive properties such as high electron-richness and excellent co-planarity of their oligomers [7], and it can be electrochemically and chemically polymerized to form stable electroactive polymers [8],[9] and [10]. The idea of solid-state polymerization of a suitable monomer in a well-ordered crystalline state was realized in the 1960 with polydiacetylenes [11]. Now, it is widely used for synthesizing the polyaniline type conducting polymers [12], [13], [14], [15] and [16]. The solid-state reaction has many advantages, such as, reduced pollution, low costs, and simplicity in process and handling. In recent years, there are a few reports regarding the solid-state synthesis of polythiophene [17], [18] and [19], and all these reports were of solid-state synthesis of polythiophene generating from 2,5-dibromothiophene derivatives as monomer. However, up to now, there was no report regarding the application of the solid-state polymerization method to synthesize the PProDOT.

In recent years, the conducting polymer/inorganic hybrid materials have been extensively investigated to obtain new kind of composite materials with synergetic or complementary behaviors, and be used in electronic or nanoelectronic devices [20], [21] and [22]. Among the inorganic materials incorporated into conducting polymers, ZnO is preferred in the preparation of hybrid materials due to its low cost, non-toxicity, good stability, high electron mobility and its extensive applications, such as photocatalyst [23] and [24]. Moreover, the electron–hole pairs can be generated when the ZnO is excited by the photons of higher than the band gap. Therefore, the electron–hole pairs can in turn react with hydroxyl ion in the reaction system to produce reactive radicals of $^\bullet O_2^-$ and $HO^\bullet$, these radicals regarding as extremely strong oxidants [25]. Furthermore, the nano-ZnO has an ideal band gap and exciton dissociation energy, and these advantages make it favorable for specific functional devices [26]. Several promising results have been reported on incorporation of nano-ZnO in polymer hybrid solar cell, transparent thin film transistors and LEDs [27], [28], [29], [30] and [31]. However, ZnO is an amphoteric oxide, and it can react with acid or base to form a water soluble salt [32]. For this reason, to successfully incorporated nano-ZnO into polymer matrix, nano-ZnO often requires surface modification. Generally, the surface modification can be realized with the low molecular surfactant, or covered with a silica layer, followed by modification with silanes [33] and [34].

In this study, we developed a solid-state method to synthesize PProDOT/ZnO composites. The content of nano-ZnO was varied from 3–7 wt% for systematically studying the effect of ZnO on the physicochemical properties of the PProDOT/ZnO composites. The structural and morphological properties of the composites were investigated by FTIR, UV–vis, X-ray diffraction and TEM. The potential application of solid-state synthesized PProDOT/ZnO composites as the electrode materials for supercapacitor were systematically evaluated by cyclic-voltammetry (CV), galvanostatic charge–discharge and electrochemical impedance spectroscopy (EIS) measurements. Furthermore, the comparative photocatalytical activity of the pure PProDOT, nano-ZnO and PProDOT/ZnO composites was investigated under UV light irradiation for the degradation of MB.

EXPERIMENTAL

Materials

3,4-Propylenedioxythiophene (ProDOT) was obtained from Aldrich, and used as received. ZnO (with an average diameter of 50 nm) and Silane Coupling Agent KH-540 (γ-Aminopropyltrimethoxysilane) were provided by Shanghai Aladdin Reagent Company, and other chemicals were used as received without further purification.

Surface Modification of Nano-ZnO

According to the literature [24], the nano-ZnO was first exposed to ambient atmosphere for 24 h to generate high-density Zn3OH groups on its surface, followed by drying at 120 °C for 2 h. Then, it was immersed in a solution of Silane Coupling Agent KH-540 (γ-Aminopropyltrimethoxysilane) in ethanol (1 g in 100 mL of ethanol) under stirring at 80 °C for 10 h, and then washed with ethanol in ultrasonic bath. Finally, the solution was filtered and dried for further use.

Synthesis of the Pprodot/Zno Composites

A typical solid-state synthesis procedure was as followed: a mixture of 0.3 g 3,4-propylenedioxythiophene (ProDOT) and 15 mg ZnO (5 wt% per monomer) in 3 mL chloroform were ultrasonicated for 30 min to facilitate monomer to adsorb on the surface of ZnO. After ultrasonication, they were placed in a vacuum oven at room temperature to evaporate the chloroform, and then the residue was transfered to the mortar, and 1.25 g anhydrous FeCl$_3$ was added. After grinding the reactants for about 1 h, the mixture became black green. The greenish black powder was washed with chloroform and ethanol, until the filtrate was colorless. Then, the powder was dried under vacuum at 60 °C for 48 h. And this sample was denoted as PProDOT/5 wt%ZnO; PProDOT/3 wt%ZnO, PProDOT/7 wt%ZnO were synthesized just by changing the amount of ZnO 3 wt% and 7 wt%, respectively.

Structure Characterization

The FTIR spectra of the samples were measured on a BRUKERQEUINOX-55 Fourier transform infrared spectrometer (Billerica, MA) at a resolution of 4 cm^{-1} using the KBr technique. UV–vis spectra of the samples were recorded on a UV–visible spectrophotometer (UV4802, Unico, USA). XRD patterns were obtained by using a Bruker AXS D8 diffractometer and the scan range (2θ) was 5°–80°, with monochromatic Cu-Ka radiation source (λ=0.15418 nm). Transmission electron microscopy (TEM) experiments were performed on a Hitachi 2600 electron microscope. The samples for TEM measurements were prepared by placing a few drops of products ethanol suspension on copper supports.

Electrochemical Tests

The working electrodes were prepared by mixing 85 wt% active materials (3 mg), 10 wt% carbon black and 5 wt% polytetrafluoroethylene (PTFE) to form slurry. The slurry was pressed on a graphite current collector with the area of 1 cm^2, and dried under vacuum at 60 °C for 24 h. All electrochemical experiments were carried out using a three-electrode system, in which the sample was used as the working electrodes, platinum as the counter electrode, saturated calomel electrode (SCE) as reference electrode, and 1 M H$_2$SO$_4$was used as electrolyte. Cyclic voltammetry (CV) and Galvanostatic charge–discharge test measurements were performed on CHI660C electrochemical working station with the potential window ranging from −0.2 V to 0.8 V. EIS was recorded by using Zennium40084 under the condition: AC voltage amplitude 5 mV, frequency range 10^5–10^{-2} Hz, and open circuit potential.

RESULTS AND DISCUSSION

FTIR Spectroscopy

Fig. 1 gives the FTIR spectra of PProDOT and PProDOT/ZnO composites. It is clear fromFig. 1 that the polymer and composites showed similar vibration bands. The bands at ~1494 and ~1320 cm^{-1} are assigned to asymmetric stretching mode of C=C and inter-ring stretching mode of C–

C, respectively. The bands at ~1173 and ~1048 cm^{-1} are associated with the C–O–C bending vibration in propylene oxide group [35] and [36]. The vibration bands at ~844 and ~657 cm^{-1} are assigned to vibrations of the C–S–C bond stretching in thiophene ring [24] and [37], while the bands at ~773 and ~434 cm^{-1} are indicated the formation of polaronic charges in thiophene [38]. In addition, the vibration band at ~1126 cm^{-1} correspond to the C–S stretching vibrations of the quinone structure originating from the thiophene ring, which suggests the successful formation of the PProDOT in this reaction [39]. These results are in good agreement with the FTIR spectra of PProDOT in previous reports [38] and [39]. The vibration bands of PProDOT/5 wt%ZnO at ~981 cm^{-1} represent the co-existence of the charge carriers of polythiophene rings [38]. And the band at ~911 cm^{-1} is due to the propylenedioxy ring deformation mode [40]. According to the previous report [41], the intensity ratio of the symmetric stretch at ~1442 cm^{-1} to the asymmetric stretch at ~1492 cm^{-1} (I_{sym}/I_{asym}) is indicative of the degree of conjugation in the polymer backbone. And I_{sym}/I_{asym} values are 0.68(PProDOT), 0.65(PProDOT/3 wt%ZnO), 0.57(PProDOT/5 wt%ZnO) and 0.64 (PProDOT/7 wt%ZnO). This suggests that the PProDOT/5 wt%ZnO composite has a longer conjugation length [42].

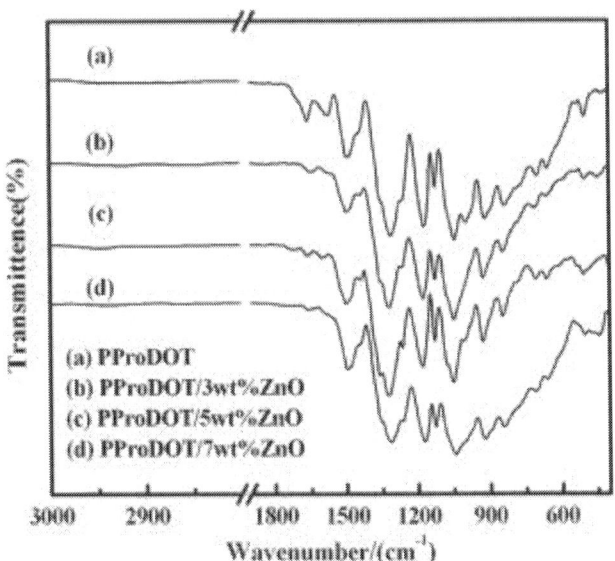

Figure 1. FTIR spectra of PProDOT and PProDOT/ZnO composites prepared with different wt% of ZnO.

UV–vis Spectroscopy

Fig. 2 shows the UV–vis spectra of PProDOT and PProDOT/ZnO composites. As can be seen from Fig. 2, there are three strong characteristic peaks for pure PProDOT: the peaks at ~425, ~455 and ~490 nm with the maximum absorption peak at ~455 nm. Generally, the peaks at ~425, ~455 and ~490 nm are assigned to the π–$\pi^{\square}$ transition of the thiophene ring [43]. Furthermore, these peaks can be considered as the absorption peaks arising from conjugated segments having different conjugation lengths [44]. In the case of PProDOT/ZnO composites, the PProDOT/3 wt%ZnO show broad absorption peak with the maximum absorption peak at ~500 nm, the PProDOT/5 wt%ZnO show two characteristic peaks located at ~460 and ~505 nm, and there are also three characteristic peaks located at ~425, ~460 and ~500 nm for PProDOT/7 wt%ZnO. Moreover, comparing with the pure PProDOT, the absorption bands corresponding to the conjugated main chains of the PProDOT/ZnO composites exhibit broad absorption peaks, and have some degree of red-shifts, which is similar to Reynolds et al. reported upon stepwise oxidation of the PProDOT-Bu$_2$[45]. These shifts occurred in composites show the increase of conjugated chain length, indicating that the presence of the nano-ZnO can enhance the conjugation degree of the polymers in the solid-state polymerization [46]. The broad absorption peaks of the composites suggest that the polymer chains are partially in doped state [8] and [47]. Therefore, it can be concluded that the PProDOT/ZnO composites have higher oxidation degree than pure PProDOT. This phenomenon can be understood by the photoelectronic characteristics of the ZnO. When ZnO is excited by the photons of higher than the band gap, electron–hole pairs are generated and in turn react with hydroxyl ion producing vary reactive radicals of $^{\bullet}O_2^{-}$ and HO$^{\bullet}$, which are extremely strong oxidants [25]. The strong oxidants stepwise enhance conjugation length and oxidative state of polymer chain. It should be noted here that the presence of the nano-ZnO content in solid-state reaction will cause the separation of the oxidant from the monomer in some degree, which, in turn, facilitates the formation of the reduction state of polymer chain during the oxidative polymerization [48]. This means that the higher amount of nano-ZnO in solid-state reaction will decrease oxidation degree of the composite. Therefore, the red-shift occurred in PProDOT/7 wt%ZnO is the lowest one among these composites.

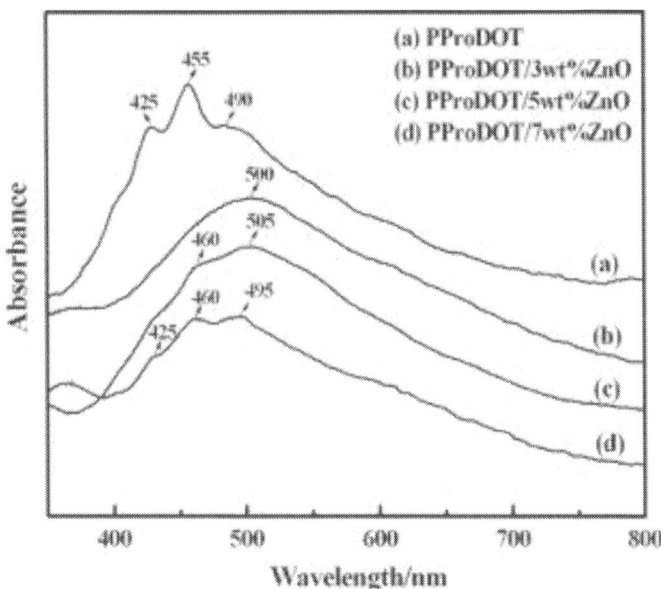

Figure 2. UV–vis spectra of PProDOT and PProDOT/ZnO composites prepared with different wt% of ZnO.

XRD Analysis

Fig. 3 shows the XRD patterns of PProDOT and PProDOT/ZnO composites. The XRD pattern of pure PProDOT shows only a characteristic peak at $2\theta=24.2°$, which associates to the intermolecular $\pi \rightarrow \pi^-$ stacking or assign to (020) reflection [45] and [49]. It can be attributed to the inter chain planar ring-stacking [50]. All the composites rather possess broad peak shape, suggesting small degree of crystallinity but amorphous structure, which is similar to other polytiophene derivative [51]. However, the addition of the nano-ZnO does not vary the crystalline structure of the forming composites. Compared to the pure ZnO, it can be seen that the XRD patterns of ZnO is not observed in composites even at a higher amount of ZnO. This situation is similar to that reported for polypyrrole/ZnO(20 wt%), in which there is no diffraction peak for ZnO even at 20 wt% ZnO in polypyrrole matrix [52]. Based on previous reports, the main reason for one cannot identify the Bragg reflections of the ZnO may be related to the negligible content of nano-ZnO left in and well embedded of nano-ZnO in polymer matrix [53], [54] and [55]. Moreover, aside from the main diffraction peaks, the sharp diffraction peaks at $2\theta=33°,35°,49°,54°$ with low intensity which are present in all composites correspond to the $FeCl_4^-$ doping agent, which generally accompanies the as made polythiophene [56].

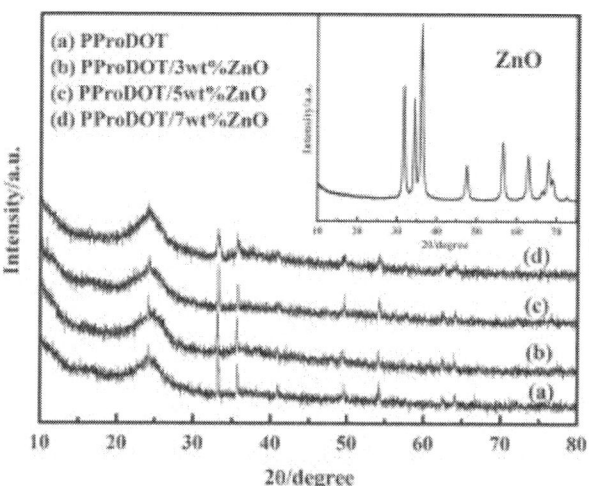

Figure 3. XRD patterns of spectra of ZnO, PProDOT and PProDOT/ZnO composites prepared with different wt% of ZnO.

TEM Studies

Fig. 4 shows the transmission electron micrograph (TEM) of the nano-ZnO, PProDOT and PProDOT/ZnO composites. The results from TEM indicate the pure nano-ZnO consists of spherical shaped particles with an average size of 50 nm (Fig. 4(a)), while all the composites with different weight ratio of nano-ZnO shows an irregular sponge like morphology. This means that the nano-ZnO has no effect on the morphology of the PProDOT. However, a close look reveals that the morphology of the pure PProDOT displays a light shaded sponge like morphology, and the PProDOT/ZnO composites exhibit some dark shaded sponge like morphology. This dark shade probably results from the ZnO entrapped by polymer matrix, implying that the nano-ZnO particles are not simply mixed up with the polymer.

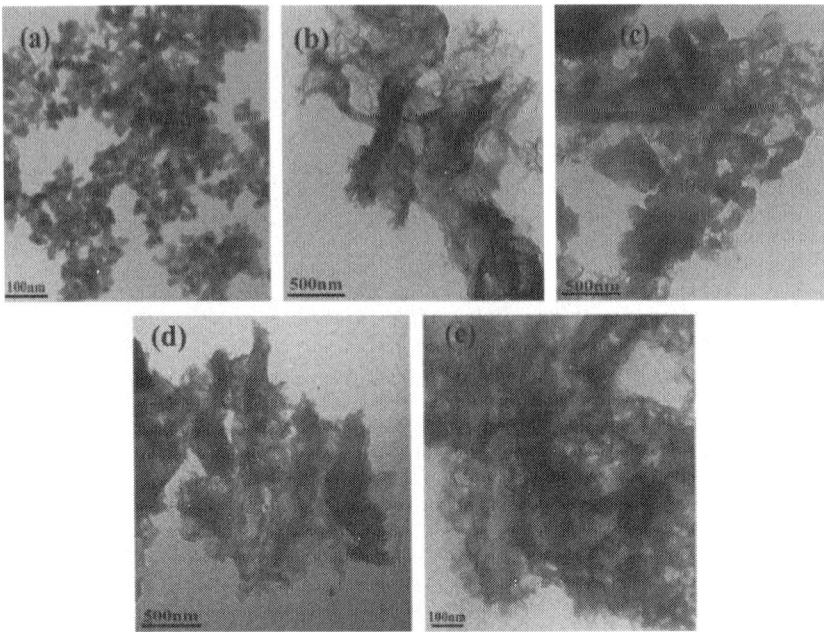

Figure 4. TEM images of the spectra of ZnO, PProDOT and PProDOT/ZnO composites prepared with different wt% of ZnO; (a) ZnO, (b) PProDOT, (c) PProDOT/3 wt%ZnO, (d) PProDOT/5 wt%ZnO and (e) PProDOT/7 wt%ZnO.

Electrical Conductivity

Cyclic voltammograms (CV) tests were performed to evaluate the capacitance properties of samples at the scan rate of 50 m V s^{-1} in 1 M H$_2$SO$_4$. The polymer and composites showed almost similar oxidation and reduction peaks. As seen in Fig. 5, the anodic (E_{pa}) and cathodic (E_{pc}) peak potentials of pure PProDOT are at 0.036–0.25 mV, 0.45–0.56 mV and, 0.13–0.07 mV, 0.31–0.44 mV, 0.62–0.73 mV. Compared with the pure PProDOT, the positive shift in the first redox peak potential and the negative shift in the first oxidation peak of the PProDOT/ZnO composites were observed. Generally, these shifts in anodic and cathodic peaks potential correspond to high charge transferring resulting from long conjugated length or higher conductivity [57]. The result confirms that the solid-state polymerized PProDOT/ZnO composites have better electrochemical activity.

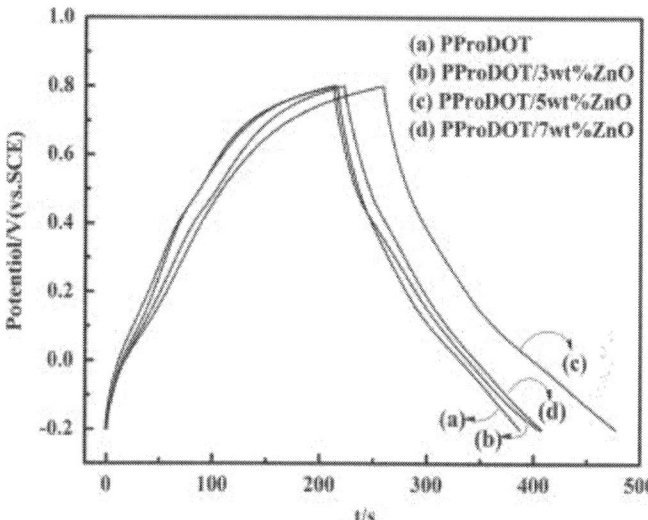

Figure 5. Cyclic voltammograms of PProDOT and PProDOT/ZnO composites prepared with different wt% of ZnO in 1 M H_2SO_4.

The galvanostatic charge–discharge measurements were also carried out at potential window between −0.2 and 0.8 V in 1 M H_2SO_4. Fig. 6 shows the galvanostatic charge–discharge of PProDOT and PProDOT/ZnO composites at a current density of 3 mA cm^{-2} in three-electrode system. The specific capacitance (SC) calculated by means of SC=$(I×\Delta t)/(\Delta V×m)$ [58], where I is charge–discharge current, Δt is the discharge time, ΔV is the electrochemical window (1V), and m is the mass of active materials within the electrode (3 mg). The specific capacitance values of PProDOT and composites calculated from Fig. 6 are 172 F g^{-1} (PProDOT),   193 F g^{-1} (PProDOT/3 wt%ZnO), 220 F g^{-1} (PProDOT/5 wt%ZnO), and 186 F g^{-1} (PPro-DOT/7 wt%ZnO), respectively. In comparison to the pure PProDOT, the composites show a higher SC and the highest SC is 220 F g^- for PProDOT/5 wt%ZnO. It should be noticed that the average specific capacitance (~190 F g^{-1}) of all the composites at the current density of 3 mA cm^{-m} is higher than previously reported poly(3,4-ethylene-dioxythiophene) [59] and [60] and polypyrrole [61] and [62]. This enhanced SC of PPro-DOT electrode can be attributed as the sponge like morphology of polymer that offers a higher specific surface area which is convenient for dopant ions accessing into the polymer matrix and inducing higher charge to keep stable. Furthermore, the SC (220 F g^{-1}) of PProDOT/5 wt%ZnO at current density of

3 mA cm^{-m} is closer to the previous literature, in which poly(3,4-ethylenedioxythiophene)/polypyrrole composite electrodes has the value of 225 F g^{-1}[63]. This enhancement of SC about the PProDOT/ZnO composites can be understood by the positive effect of the nano-ZnO during the solid-state polymerization, such as high electron mobility. Moreover, it is concluded from the FTIR spectra and UV–vis absorption spectra that the composites with the presence of ZnO varying from 3 wt% to 7 wt% have higher conjugation degree and doping level than pure PProDOT, which in turn enhances the electro chemical activities of the composite electrodes. PProDOT/5 wt%ZnO has the highest SC among the composites, which may be related to the higher conjugation degree of polymer chain.

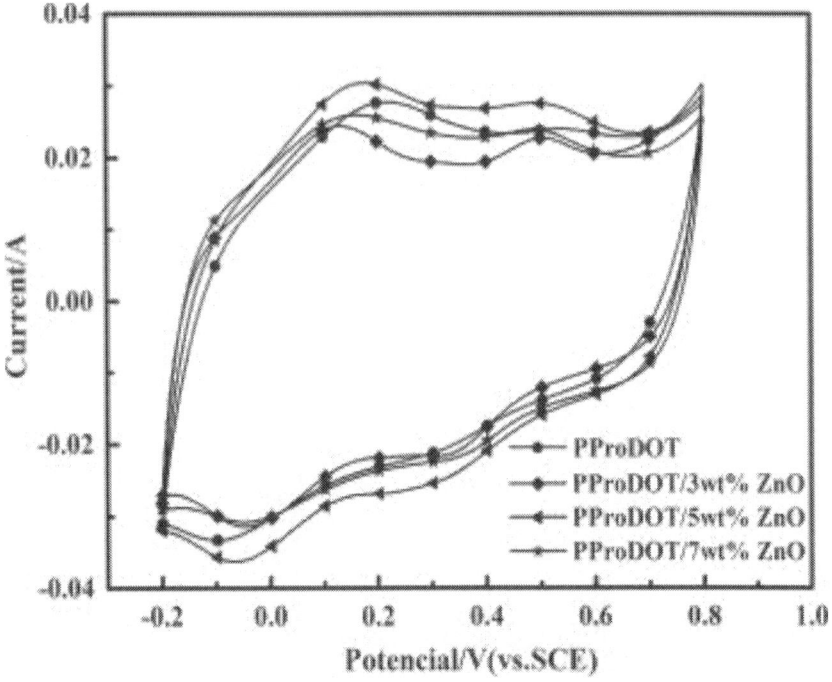

Figure 6. Galvanostatic charge–discharge curves of PProDOT and PProDOT/ZnO composites prepared with different wt% of ZnO at 3 mA cm^{-2}, mass of 3 mg, three-electrode system, and electrolyte of $1 \text{ M H}_2\text{SO}_4$.

Electrochemical impedance spectroscopy (EIS) was used to study the resistance limitation of electrically conducting polymers. The EIS data were analyzed using Nyquist plots, which show the frequency response of the supercapacitors assembly and the imaginary part (Z'') of the impedance against the real part (Z'). Nyquist plots of PProDOT and PProDOT/ZnO composites at 5 mV over the frequency range of 0.01 Hz–100 KHz are given in Fig. 7. It can be seen that EIS plots contain two well separated patterns. Firstly, the high frequency intercept of the semi-circle with the real axis can be used to evaluate the value of internal resistance, which included the resistance of the electrolyte solution, the intrinsic resistance of the active material, and the contact resistance at the interface active material/current collector. The values of internal resistance obtained from Fig. 7 are 0.32 Ω (PProDOT), 0.44 Ω (PProDOT /3 wt%ZnO), 0.45 Ω (PProDOT/5 wt%ZnO), and 0.38 Ω (PProDOT/7 wt%ZnO). The radius of the semicircular represents the charge transfer resistance. Therefore, the charge transfer resistance obtained from Fig. 7 are 0.42 Ω (PProDOT), 0.31 Ω (PProDOT/3 wt%ZnO), 0.29 Ω (PProDOT/5 wt%ZnO), and 0.32 Ω (PProDOT / 7 wt% ZnO). The small semicircle might be due to diffusion effect of the electrolyte in the electrodes [64]. Secondly, at low frequencies, the vertical line indicates the pure capacitive behavior, the more vertical curve suggests the better capacitive behavior of supercapacitor [65]. It can be clearly seen that the inclined line with a slope is more close to 90° for the PProDOT/5 wt%ZnO, which was a characteristic feature of pure capacitive behavior [66]. While the slope is close to 45° for pure PProDOT, indicating the existence of Warburge resistance resulting from the frequency dependence of ion diffusion in the electrolyte to the electrode interface appearing in the higher voltage. These results also further illustrate that PProDOT/5 wt%ZnO can be used as electrode material for super capacitors.

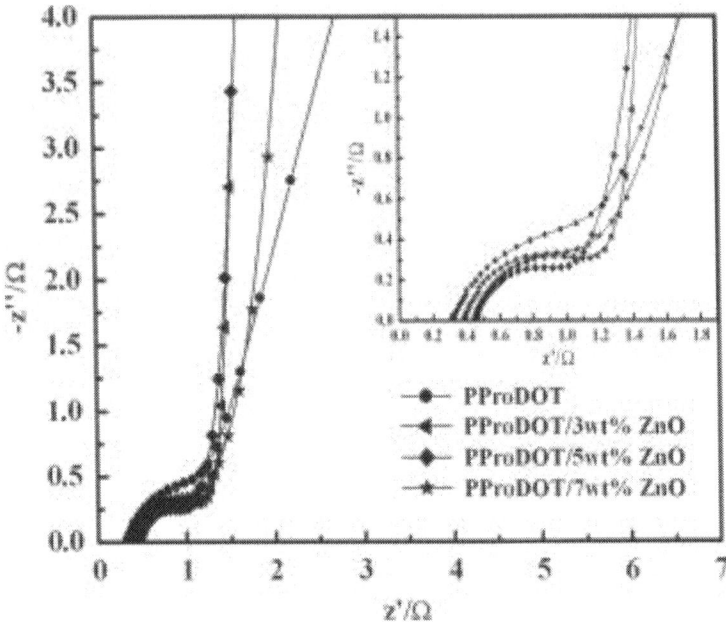

Figure 7. EIS curves of PProDOT and PProDOT/ZnO composites prepared with different wt% of ZnO at open-circuit potential, 0.5 V with amplitude of 5 mV over the frequency range of 0.01 Hz–100 KHz.

Photocatalytic Activity

The photocatalytic degradation of MB dyes in the presence of the PProDOT/ZnO composite as catalyst under UV light source (λ=365 nm) at different irradiations time was investigated. According to the previous report [67], we used 0.4 mg/mL catalyst and 1×10^{-0} M MB dye solution for study. As it is shown in Fig. 8, the decrease of the absorption band intensities of the dyes indicates that the dyes have been efficiently degraded by PProDOT/ZnO composites photocatalyst. When the MB dye solution was exposed to the UV light for 5 h, the degradation efficiencies of the MB dye are 42% (PProDOT/3 wt%ZnO), 52% (PProDOT/5 wt%ZnO) and 38% (PProDOT/7 wt%ZnO) photocatalysts. For comparison, the photocatalytic degradation of MB dyes in the presence of the nano-ZnO and the pure PProDOT was also investigated, and the degradation efficiency of the MB dye is 31% and 14%, respectively. The results in Fig. 8(d) indicated that no significant changes were observed in absorption spectra of dyes by the pure PProDOT hotocatalysts. The result showed that the photocatalytic activity of the pure PProDOT is negligible. Furthermore, Fig. 8(f) shows the variation in degradation efficiency versus irradiation time for the MB dyes solutions in the presence of the ZnO nanoparticles, pure PProDOT and the PProDOT/ZnO composites hotocatalysts. After these investigations, it is clear that the PProDOT/5 wt%ZnO is more efficient under UV light conditions for the degradation of the MB dyes. The higher conjugation and oxidation degree of the PProDOT/5 wt%ZnO due to the good electron mobility enhance its photocatalytic activity. Therefore, the photocatalytic behavior of the PProDOT/5 wt%ZnO may result from the synergetic effect between PProDOT and ZnO.

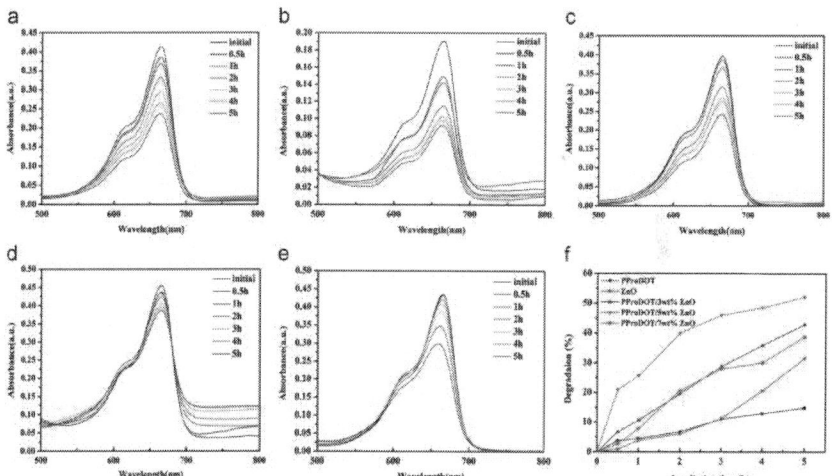

Figure 8. UV–vis absorption spectra of MB dyes by PProDOT/ZnO composites for different irradiation times under UV light irradiation (a) PProDOT/3 wt%ZnO, (b) PProDOT/5 wt%ZnO, (c) PProDOT/7 wt%ZnO, (d) PProDOT, (e) Nano-ZnO, and (f) degradation efficiency of the MB dyes.

CONCLUSION

In this paper, the poly(3,4-propylenedioxythiophene)/Zinc Oxide composites (PProDOT/ZnO) were synthesized by the solid-state oxidative polymerization method. The results showed that the ZnO may bring a higher conjugation length and more oxidative state to the polymer chains during the solid-state polymerization, which was related to the formation of reactive radicals of $^{\bullet}O_2^{-}$ and $HO^{\bullet}$. XRD and morphological analyses indicated that the negligible content of nano-ZnO left in polymer matrix, and nano-ZnO were well imbedded in polymer matrix. In addition, the sponge like morphology of polymers offered a higher specific capacitance, which was resulted from higher surface area that was convenient for dopant ions accessing into the polymer matrix and inducing higher charge to keep stable. Moreover, the composites also had good photocatalytic activity on MB dyes under UV light irradiation.

ACKNOWLEDGMENTS

We gratefully acknowledge the financial support from the National Natural Science Foundation of China (Nos. 21064007, 21264014) and Opening Project of Xinjiang Laboratory of Petroleum and Gas Fine Chemicals (XJDX0908-2011-05).

REFERENCES

1. J.C. Chiang, A.G. MacDiarmid, Synth. Met. 13 (1986) 193–205.
2. A.F. Diaz, J.I. Castillo, J. Logan, W.Y. Lee, J. Electroanal. Chem. Interfacial Electrochem. 129 (1981) 115–132.
3. R.J. Waltman, J. Bargon, A. Diaz, J. Phys. Chem. 87 (1983) 1459– 1463.
4. Q. Pei, G. Zuccarello, M. Ahlskog, O. Inganäs, Polymer 35 (1994) 1347–1351.
5. A. Kumar, R. Singh, S.P. Gopinathan, A. Kumar, Chem. Commun. 48 (2012) 4905–4907.
6. T. Dey, M.A. Invernale, Y. Ding, Z. Buyukmumcu, G.A. Sotzing, Macromolecules 44 (2011) 2415–2417.
7. Y. Liang, B. Peng, J. Liang, Z. Tao, J. Chen, Org. Lett. 12 (2010) 1204–1207.
8. M.R. Rosario-Canales, P. Deria, M.J. Therien, J.J. Santiago-Avilés, ACS Appl. Mater. Interfaces 4 (2012) 102–109.
9. D.M. Welsh, A. Kumar, E. Meijer, J. Reynolds, Adv. Mater. 11 (1999) 1379–1382.
10. F. Miomandre, P. Audebert, K. Zong, J.R. Reynolds, Langmuir 19 (2003) 8894–8898.

11. K. Yee, R. Chance, J. Polym. Sci.: Polym. Phys. Ed. 16 (2003) 431–441.
12. J. Gong, X.J. Cui, Z.W. Xie, S.G. Wang, L.Y. Qu, Synth. Met. 129 (2002) 187–192.
13. J. Stejskal, A. Riede, D. Hlavatá, J. Prokeš, M. Helmstedt, P. Holler, Synth. Met. 96 (1998) 55–61.
14. T. Abdiryim, Z. Xiao-Gang, R. Jamal, Mater. Chem. Phys. 90 (2005) 367–372.
15. J. Huang, J.A. Moore, J.H. Acquaye, R.B. Kaner, Macromolecules 38 (2005) 317–321.
16. X.S. Du, C.F. Zhou, G.T. Wang, Y.W. Mai, Chem. Mater. 20 (2008) 3806–3808.
17. H. Meng, D.F. Perepichka, F. Wudl, Angew. Chem. Int. Ed. 42 (2003) 658–661.
18. H.J. Spencer, R. Berridge, D.J. Crouch, S.P. Wright, M. Giles, I. McCulloch, S.J. Coles, M.B. Hursthouse, P.J. Skabara, J. Mater. Chem. 13 (2003) 2075–2077.
19. H. Meng, D.F. Perepichka, M. Bendikov, F. Wudl, G.Z. Pan, W. Yu, W. Dong, S. Brown, J. Am. Chem. Soc. 125 (2003) 15151–15162.
20. K. Gurunathan, D. Amalnerkar, D. Trivedi, Mater. Lett. 57 (2003) 1642–1648.
21. V. Khomenko, E. Frackowiak, F. Beguin, Electrochim. Acta 50 (2005) 2499–2506.
22. J. Yan, T. Wei, B. Shao, Z. Fan, W. Qian, M. Zhang, F. Wei, Carbon 48 (2010) 487–493.
23. I. Musa, F. Massuyeau, E. Faulques, T.P. Nguyen, Synth. Met. 162 (2012) 1756–1761.
24. X. Chen, Z. Zhou, W. Lv, T. Huang, S. Hu, Mater. Chem. Phys. 115 (2009) 258–262.
25. D. Fu, G. Han, Y. Chang, J. Dong, Mater. Chem. Phys. 132 (2011) 673–681.
26. A. Chaurasia, L. Wang, L.H. Gan, T. Mei, Y. Li, Y.N. Liang, X. Hu, Eur. Polym. J. 49 (2012) 630–636.
27. S.D. Oosterhout, M.M. Wienk, S.S. van Bavel, R. Thiedmann, L.J.A. Koster, J. Gilot, J. Loos, V. Schmidt, R.A.J. Janssen, Nat. Mater. 8 (2009) 818–824.
28. B.S. Ong, C. Li, Y. Li, Y. Wu, R. Loutfy, J. Am. Chem. Soc. 129 (2007) 2750–2751.
29. D. Guo, C. Wu, H. Jiang, Q. Li, X. Wang, B. Chen, J. Photochem. Photobiol. B: Biol. 93 (2008) 119–126.
30. J.H. Lim, C.K. Kang, K.K. Kim, I.K. Park, D.K. Hwang, S.J. Park, Adv. Mater. 18 (2006) 2720–2724.
31. J. Dai, C.X. Xu, X.W. Sun, Adv. Mater. 23 (2011) 4115–4119.
32. Z.h. Yuan, J.h. Jia, L.d. Zhang, Mater. Chem. Phys. 73 (2002) 323–326.

33. C.H. Lu, C.H. Yeh, Mater. Lett. 33 (1997) 129–132.
34. H. Li, Y. Chen, C. Ruan, W. Gao, Y. Xie, J. Nanopart. Res. 3 (2001) 157–160.
35. Y. Yang, Y. Jiang, J. Xu, J. Yu, Polymer 48 (2007) 4459–4465.
36. Y. Soo Kim, S. Bin Oh, J. Hyeok Park, M. Suk Cho, Y. Lee, Sol. Energy Mater. Sol. Cells 94 (2010) 471–477.
37. K.R. Reddy, W. Park, B.C. Sin, J. Noh, Y. Lee, J. Colloid Interface Sci. 335 (2009) 34–39.
38. S. Sindhu, C. Siju, S. Sharma, K. Rao, E. Gopal, Bull. Mater. Sci. 35 (2012) 611–616.
39. N.A. Kumar, H.J. Choi, A. Bund, J.B. Baek, Y.T. Jeong, J. Mater. Chem. 22 (2012) 12268.
40. L. Zhan, Z. Song, J. Zhang, J. Tang, H. Zhan, Y. Zhou, C. Zhan, Electrochim. Acta 53 (2008) 8319–8323.
41. C. Wang, M.E. Benz, E. LeGoff, J.L. Schindler, J. Allbritton-Thomas, C.R. Kannewurf, M.G. Kanatzidis, Chem. Mater. 6 (1994) 401–411.
42. Y. Furukawa, M. Akimoto, I. Harada, Synth. Met. 18 (1987) 151–156.
43. T. Yamamoto, T. Shimizu, E. Kurokawa, React. Funct. Polym. 43 (2000) 79–84.
44. J.J. Apperloo, R. Janssen, M.M. Nielsen, K. Bechgaard, Adv. Mater. 12 (2000) 1594–1597.
45. D.M. Welsh, L.J. Kloeppner, L. Madrigal, M.R. Pinto, B.C. Thompson, K.S. Schanze, K.A. Abboud, D. Powell, J.R. Reynolds, Macromolecules 35 (2002) 6517–6525.
46. E Eren, G. Celik, A. Uygun, J. Tabačiarová, M. Omastová, Synth. Met. 162 (2012) 1451–1458.
47. A. Balamurugan, K.C. Ho, S.M. Chen, Synth. Met. 159 (2009) 2544– 2549.
48. T. Abdiryim, A. Ubul, R. Jamal, Y. Tian, T. Awut, I. Nurulla, J. Appl. Polym. Sci. 126 (2012) 697–705. 530 A. Ali et al.
49. T.Y. Kim, C.M. Park, J.E. Kim, K.S. Suh, Synth. Met. 149 (2005) 169–174.
50. C. Jiang, G. Chen, X. Wang, Synth. Met. 162 (2012) 1968–1971.
51. P. Vacca, G. Nenna, R. Miscioscia, D. Palumbo, C. Minarini, D.D. Sala, J. Phys. Chem. C 113 (2009) 5777–5783.
52. M. Chougule, S. Sen, V. Patil, J. Appl. Polym. Sci. 125 (2012) 541–547.
53. K.G.B. Alves, J.F. Felix, E.F. de Melo, C.G. dos Santos, C.A.S. Andrade, C.P. de Melo, J. Appl. Polym. Sci. 125 (2012) 141–147.
54. A. Mostafaei, A. Zolriasatein, Prog. Nat. Sci.: Mater. Int. 22 (2012) 273–280.
55. Y. Chen, Z. Zhao, C. Zhang, Synth. Met. 163 (2013) 51–56.
56. T. Abdiryim, R. Jamal, C. Zhao, T. Awut, I. Nurulla, Synth. Met. 160 (2010) 325–332.
57. S. An, T. Abdiryim, Y. Ding, I. Nurulla, Mater. Lett. 62 (2008) 935–938.

58. T. Abdiryim, A. Ubul, R. Jamal, F. Xu, A. Rahman, Synth. Met. 162 (2012) 1604–1608.
59. D. Antiohos, G. Folkes, P. Sherrell, S. Ashraf, G.G. Wallace, P. Aitchison, A.T. Harris, J. Chen, A.I. Minett, J. Mater. Chem. 21 (2011) 15987–15994.
60. K. Lota, V. Khomenko, E. Frackowiak, J. Phys. Chem. Solids 65 (2004) 295–301.
61. E. Frackowiak, K. Jurewicz, K. Szostak, S. Delpeux, F. Beguin, Fuel Process. Technol. 77 (2002) 213–219.
62. Y.m. Cai, Z.y. Qin, L. Chen, Prog. Nat. Sci.: Mater. Int. 21 (2011) 460–466.
63. Y. Xu, J. Wang, W. Sun, S. Wang, J. Power Sources 159 (2006) 370–373.
64. S.R.P. Gnanakan, M. Rajasekhar, A. Subramania, Int. J. Electrochem. Sci. 4 (2009) 1289–1301.
65. H. Mi, X. Zhang, X. Ye, S. Yang, J. Power Sources 176 (2008) 403–409.
66. Z.J. Li, T.X. Chang, G.Q. Yun, Y. Jia, Powder Technol. 224 (2012) 306–310.
67. V. Eskizeybek, F. Sarı, H. Gülce, A. Gülce, A. Avcı, Appl. Catal. B: Environ. 119–120 (2012) 197–206.

CITATION

Ahmat Ali, Ruxangul Jamal, Weiwei shao, Adalet Rahman, Yakupjan Osman, Tursun Abdiryim, Structure and properties of solid-state synthesized poly(3,4-propylenedioxythiophene)/nano-ZnO composite, Progress in Natural Science: Materials International, Volume 23, Issue 6, December 2013, Pages 524-531, ISSN 1002-0071, http://dx.doi.org/10.1016/j.pnsc.2013.11.002.

CHAPTER 9

Spintronics Driven by Superconducting Proximity Effect

Guoxing Miao

Institute for Quantum Computing and Electrical and Computer Engineering, University of Waterloo, Waterloo ON, Canada

INTRODUCTION

In this chapter, we will discuss a few selected topics on the applications of superconducting proximity effect, and the related inverse proximity effect, in the field of spintronics. Superconducting proximity effect occurs when Cooper pairs from a superconductor propagate into the adjacent metallic systems and induce superconducting correlations in the otherwise non-superconducting materials. Due to the limited coherence length of Cooper pairs, this effect is confined to the very interface between the two materials and can be used as a method to trigger supercurrent flow, or to create particle-hole symmetry, in a wide range of devices. The inverse proximity effect can be viewed as the counteraction of the above-mentioned effect. Superconductivity in the original superconductor material inevitably weakens when it drives superconductivity into its neighbours; in addition, back flow of unpaired electrons and sometimes spin polarized electrons, will create more pair breaking within the superconductor and weaken it further. These effects can, however, be used as effective ways to control superconductivity through spin manipulations.

This chapter is organized in the following way. After a brief introduction on superconducting proximity effect and inverse proximity effects, we will continue our discussions from a few device points of view. 1. Proximity induced superconductivity in low dimensional electron systems, such as in the surface states of a 3D topological insulator and in the Rashba-split bands of a heavy metal/semiconductor nanowire. These form the most promising platforms that can host the elusive Majorana fermions for quantum computing applications. 2. Inverse proximity from multiple ferromagnetic neighbours can controllably turn superconductivity On and Off. The change in superconductivity states also leads to a large change in the device resistance, known as the superconducting spin valve effect.

- Superconducting proximity effect — the example of Majorana fermion creation
- Inverse proximity on superconductors — the example of superconducting spin valves

SUPERCONDUCTING PROXIMITY EFFECT — THE EXAMPLE OF MAJORANA FERMION CREATION

The term superconducting proximity effect [1, 2], by default, refers to the leak-out of a superconductor's Cooper pair wave function into an otherwise non-superconducting material, and the generation of induced superconductivity in that material. This is a spontaneous process that occurs whenever the superconductor makes clean enough contact with a metallic system. Fig.1 illustrates the distribution of superconductor pairing potential Δ across the interface between a superconductor and a normal metal. Far away from the interface, Δ simply sits at the respective bulk values, being Δ_0 on the superconductor side and vanishing on the normal metal side. The proximity effect shows up clearly near the interface: Δ is weakened on the superconductor side but gradually emerges in the normal metal towards the interface. Naturally, the two systems have to be in atomic contact for such wave function overlap to happen effectively. The interface transparency is critical for the strength of the proximity, and is often characterized with an impedance parameter γB such that the interfacial transparency $T=11+\gamma B$. The transparency captures the interfacial quality as well as the wave function matching between the materials. In Fig.1, a clear discontinuity is seen across the interface which indicates a less-than-perfect interfacial transparency. A cleaner interface

means stronger proximity, which adds some quite stringent requirements
to the materials' development.

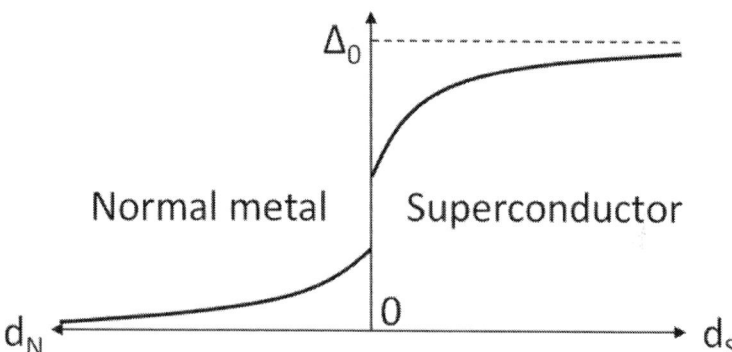

Figure 1. Evolution of the superconductor pairing potential Δ near an S/N
interface. Δ_0 is the bulk pairing potential of the superconductor.

Such superconducting proximity effect is of particular interest to the
spintronics community because it not only allows for spontaneously
driving a non-superconducting material into superconductivity, but also
enables actively pumping supercurrents over a prolonged distance across a
non-superconducting material. Their fundamental principles are the same:
extension of the superconductor wave function into the non-
superconducting material. Naturally, the proximity effect can propagate
into a material as long as there are available electronic states near the
Fermi level: i.e., a metallic material. There have been some quite
important reports in the field demonstrating superconductor coupling
through unconventional media: for example, the manifestation in 2D
semiconductor electron gases [3, 4, 5] and topological insulator surface
states [6, 7, 8]; the two-valley nature with relativistic Dirac electrons of
graphene [9]; the gate tunability when coupled to 1D carbon nanotubes
[10, 11], as well as with 0D C_{60} molecules [12] and InAs quantum dots
[13]; the flow of supercurrents through double-stranded DNA molecules
[14]; and the conversion into spin triplet supercurrents in half-metals
[15, 16].

Whereas the superconducting proximity effect can essentially drive any
electron system into superconducting, we focus our attention on a very

specific topic: generation of the long-sought-after Majorana fermions on superconductor platforms and application of these particles to topological quantum information processing [17]. A Majorana fermion, by definition, is a spin-1/2 particle identical to its own antiparticle, as opposed to a conventional Dirac fermion. Being its own antiparticle suggests that the particle has to be charge neutral as well. A superconductor's wavefunction has perfect electron-hole symmetry, making it an ideal platform for creating objects with similar properties. Other than being electrically charged, the quasiparticle excitations in a superconductor come really close to the Majorana fermions. Naturally, many of the proposals for creating Majorana fermions in solid-state systems involve superconductivity as one of the core ingredients. There are a number of ingenious device concepts for realizing Majorana fermions in a superconductor platform. The most practical route of synthesizing a localized Majorana fermion is to build on the platform of p-wave superconductors [18, 19] such as strontium ruthenate (Sr_2RuO_4): a spinless p_x+ip_y superconductor [20]. A p-wave superconductor's wavefunction is two-fold degenerate and has chirality on the angular momentum; it is therefore often termed the chiral p-wave state. It differs from the more conventional s-wave BCS superconductors and d-wave high-temperature superconductors in the sense that the electrons prefer to pair into spin-triplet configurations. Since ferromagnetism is observed in many closely related strontium ruthenate compounds (such as $SrRuO_3$), it is natural to consider the possibility of triplet pairing on these superconducting species. There has been compelling evidence confirming that Sr_2RuO_4 is indeed a p-wave superconductor [21]. This spontaneous time-reversal symmetry breaking ensures that the system is stable against magnetic impurity scattering, but not so much against normal elastic impurity scatterings, such as from non-magnetic defects, grain boundaries, or surfaces. As a result, it is very hard to obtain materials with a high enough quality for device purposes. With Sr_2RuO_4 as an example, while its T_C can reach as much as 1.5 K in the best-quality single crystals, it is practically non-superconducting in any of the thin films deposited so far [21], which is a significant obstacle for its potential application in the topological quantum computer architectures.

The superconductor proximity effect comes in as an experimentally feasible route to "simulate" a p-wave superconductor that can subsequently host Majorana modes on the resulting 2D platforms [22,23]. Now that the particle-hole symmetry has already been guaranteed on the given platforms by the superconductivity induced from the superconducting proximity effect, what about the spinless nature of the p-wave superconductors? The signature we need to look for is the so-called helical Dirac spin state, as illustrated in Fig.2, which readily shows up on the surface of a 3D topological insulator [24]. A 3D topological insulator is a quantum spin Hall insulator with an inside that is insulating and a surface that is conducting [25]. Due to the nontrivial topology of the material, its interface with a topologically trivial insulator, such as a normal insulator or vacuum, always has the conducting surface states present, which are strictly protected by the topological orders. We will not go into too much detail about topological insulator itself as it is not the focus of this article, except to point out that: a) the electron dispersion relation shows a helical spin texture — the electrons' spin orientation and momentum direction are tightly locked due to strong spin-orbit coupling; b) the dispersion is linear for the energy range we are interested in — known as the relativistic Dirac dispersion; and c) there is no energy gap and no spin degeneracy — a single band structure, different from a zero-gap semimetal or the spin and valley degenerate graphene. As the ingenious proposal from Fu and Kane [22] demonstrates, by coupling the helical surface states of a topological insulator with a conventional s-wave BCS superconductor through proximity effect, one can essentially simulate the properties of a p-wave superconductor and perform any subsequent operations with the composite states. Due to the two-dimensional nature of the topological surface states, the resultant "p-wave superconductor" is also two-dimensional. When a vortex nucleates on such a surface, it defines a non-superconducting region which essentially creates an "edge" on the surface localized to a tiny circle. A closed contour around this circle has a Berry phase shift of π due to the rotation of spin orientation around this circle. A Majorana zero mode is trapped within this vortex core, and can therefore be moved coherently when one physically moves this vortex around, as if it is a standalone physical particle. Once these localized, non-Abelian Majorana bound states are established, one can then perform the necessary braiding operation on them which covers all the quantum gates required for quantum information processing [17].

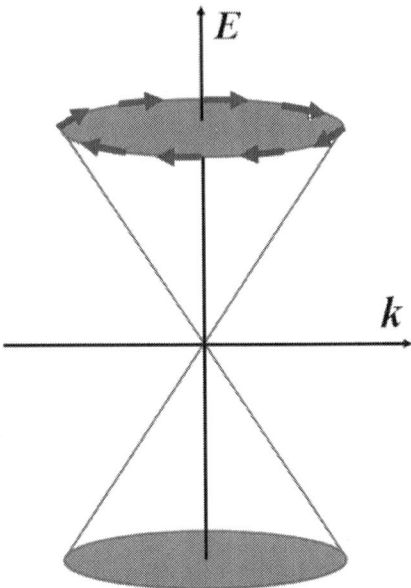

Figure 2. Electron energy dispersion with a helical spin texture similar to the surface states of a topological insulator. Blue arrows indicate the directions of the spins, and red cones indicate the linear dispersion near the Dirac point.

Though a topological insulator makes a convenient platform for creating a p-wave superconductor and the subsequent Majorana fermions, the above proposal remains an experimental challenge mainly due to the difficulties associated with perfecting the topological materials themselves, which requires significantly more effort along the road [24, 26, 27, 28]. Experimentalists have progressed remarkably further on a much more mature material platform: the surface of a semiconductor with strong Rashba spin-orbit coupling. Evidence of Majorana bound states has already been reported in this direction [29, 30, 31]. The basic idea here is to also look for an electronic state that resembles the helical spin states as described above, and the Rashba spin-orbit coupling, resulting from structural change at interfaces; this idea can become a convenient solution to the problem. Fig.3 illustrates the effect of Rashba spin-orbit coupling on a "free" electron band. Due to the relativistic effect, electrons moving in the interfacial electric fields also experience effective magnetic fields on their rest frame. Forward-moving electrons and backward-moving electrons see opposite in-plane effective magnetic fields. In the forward-moving direction, the spin-"inward" electrons are lifted higher in energy than the spin-"outward" electrons, and vice versa in the opposite-moving direction. Here the inward and outward spin orientations are with respect to the paper of drawing, and which spin configuration has lower energy is also determined by the Rashba field direction of that interface. Note that the spins are still degenerate at the k=0 point where the spin-orbit coupling vanishes, and it creates a crossing between the two bands. The dispersion relation is nearly linear in k close to this crossing point. Fig.3only depicts the situation of one-dimensional electron motions (x-direction). Imagine that we are actually dealing with a two-dimensional surface, and the k space extends in both x and y directions. This will revolve the spin-split bands around the centre axis of k=0, and the crossing point evolves into a Dirac point under the electrons' two-dimensional motion. Because the Rashba field is normal to the xy-plane electron motion, the resulting effective magnetic fields — and therefore the principle spin quantum axes —always lie within the xy-plane and transverse to the direction of motion. This leads to a helical spin texture as depicted in Fig.3. Note that two helical spin-textured Fermi surfaces exist in this structure and a perpendicular Zeeman field along z-direction is necessary to open a gap at the Dirac point and remove the spin degeneracy from this system, eventually rendering it spinless. With this construction of a helical spin state through Rashba spin-orbit coupling, we can again proceed with superconducting proximity effect to simulate the properties of a p-wave superconductor.

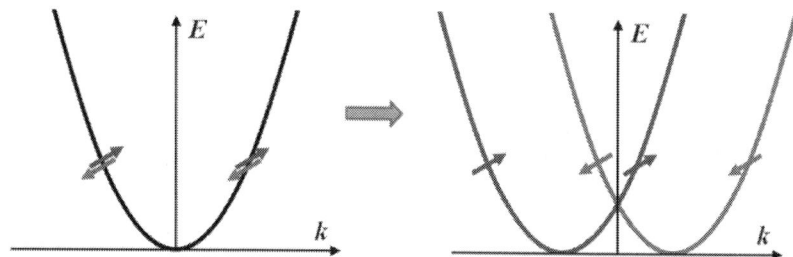

Figure 3. Illustration of a Rashba type spin-orbit splitting. Left: a typical parabolic "free" electron band with two-fold spin degeneracy. Right: In the presence of Rashba spin-orbit coupling, moving electrons feel an effective magnetic field on their rest frame, and the two spin channels are therefore split in energy. The effect is linear with respect to the wave vector k, and opposite for the two types of spins. Therefore, the Rashba spin-orbit coupling adds a linear component to the quadratic function, creating a uniform shift of the parabolas from the origin but in opposite directions for the opposite spin channels.

The proposal from Das Sarma et al. [23] is based on a 2D semiconductor electron system governed by the following equation of motion,

$$H_0 = \frac{\hbar^2 p^2}{2m^*} - \mu + V_z \sigma_z + \alpha \left(\vec{\sigma} \times \vec{p} \right) \cdot \hat{z}$$

$$(1)$$

Here, m^*, μ, V_z, α, are the electrons' effective mass, the chemical potential, the out-of-plane Zeeman splitting, and the Rashba spin-orbit coupling strength, respectively. Without the Zeeman term, the band structure would look just like that in Fig.3: revolving around the centre axis, as described above. We can easily see that the Zeeman term has a negligible influence on the total wavefunction when $\vec{p}$ is large, but becomes dominant close to $p \rightarrow =0$ and opens up an energy gap right at the Dirac point, with the gap size $2V_z$ determined by the electrons' g-factor and the applied Zeeman field strength. This perpendicular Zeeman field is critical in this proposal for removing the spin degeneracy from the structure, leaving only one band present within the Zeeman gap. One does need to carefully tune the Fermi level of the system to within this gap to take advantage of this feature. The next ingredient is the superconducting proximity effect, which will induce superconductivity into the semiconductor's Rashba-split surface states. In order for the proximity effect to be effective, the semiconductor needs to be sufficiently conducting, in addition to possessing strong Rashba spin-orbit coupling. These requirements readily identify some of the heavy-element, low-gap semiconductors as the most promising choices, such as InSb or InAs. They tend to also have very large conduction electron g-factors, dozens of times larger than 2, making them even more ideal in applied magnetic fields. The superconductors, on the other hand, tend to have lower critical magnetic fields if the field direction is pointing out of the film plane, and in many cases the superconductivity will be completely destroyed by a field of a few Tesla. In order to induce appreciable Zeeman splitting, hopefully a few meV or higher, without suppressing the superconductivity, magnetic insulators are often used in these proposals to induce the necessary perpendicular Zeeman field. These materials are known to produce large exchange Zeeman fields onto adjacent free electrons, and are most effective on low-dimensional 2D or 1D electron systems [32]. The induced effective field can reach hundreds of Tesla with this approach. With exchange fields being short ranged, they have negligible influence on the superconductivity because the superconductor and the magnetic insulator are not in direct contact. Later, Alicea et al. [33] offered a clever twist on the above proposals by combining the Dresselhaus spin-orbit coupling, intrinsic to III-V semiconductors because of their crystallographic inversion asymmetry, and the Rashba spin-orbit coupling, due to the structural inversion asymmetry at interfaces, in a semiconductor quantum well structure with selected crystalline orientations. Now that the principle spin quantum axis no longer lies within the electrons' motion plane, a magnetic field with reasonable strength applied parallel to the plane surface, coupled with the relatively large g-factors in these semiconductors, can also open up an appreciable energy gap at the centre point, removing the spin degeneracy found there. Once the spinless platform is constructed, the next important practice is to tune the chemical potential of the system right to the middle of this gap through chemical doping or electric field gating. One can then place a conventional s-wave superconductor on top, which induces superconductivity into the system through proximity effect. The induced superconducting correlation opens up another energy gap on the conduction electrons and the system becomes fully gapped. With all these ingredients in place, the system can be described with exactly the same Bogoliubov-de Gennes (BdG) equations as the topological p-wave superconductors, and Majorana bound states are expected to exist at vortex cores on this special platform.

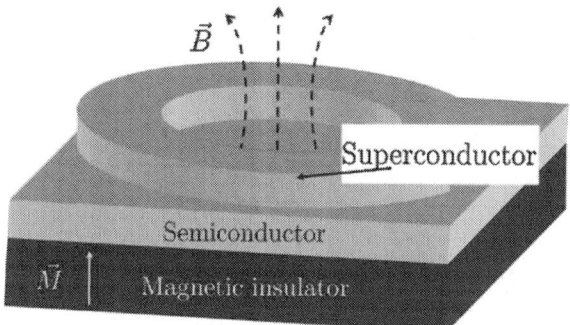

Figure 4. Majorana state bound to a vortex core on a heterostructure of a Rashba semiconductor platform, with both superconducting and magnetic proximity effects [23].

Another route of synthesizing Majorana fermions is to perform superconducting proximity on top of a heavy metal surface with strong Rashba spin-orbit coupling [34, 35]. The Rashba splitting tends to be much larger than in the semiconductor systems. Under the strong Rashba spin-orbit coupling, the metal's surface states also show two spin-split parabolas, with a crossing at k=0 similar to those shown in Fig.3. A perpendicular external magnetic field, or a perpendicular exchange field from magnetic insulators, will open up a Zeeman gap at the cross point, and an s-wave superconductor will induce superconductivity into the system through superconducting proximity effect. While the superconducting gap stays with the chemical potential of the system, one still needs to match the chemical potential to the middle of the Zeeman gap. Due to the abundance of carriers in the metallic systems, however, it is extremely difficult to move the chemical potential through electric field gating. Therefore, one has to carefully select a suitable electron system to begin with, with the crossing point of the surface states not too far from the metal's Fermi level. Fig.5 shows one such set-up. Tunnelling spectrum to the interface can reveal the existence of the desired, gapped Rashba surface states. Once the superconducting proximity effect kicks in, appearance of Majorana zero-energy modes can mediate resonant Andreev reflection between two leads, with perfect conductance of $2e^2/h$ independent on the coupling strength. As a comparison, conventional resonant Andreev reflection only shows the maximum conductance when the two coupling amplitudes are identical.

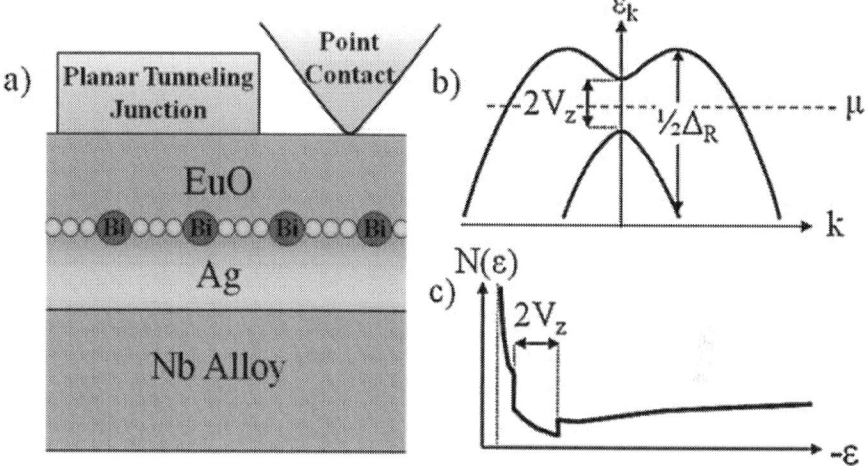

Figure 5. The proposed sample layout for the heavy-metal-based Majorana fermion creation [34]. (a) Ag(111) surface with 1/3 monolayer of Bi coverage in the $(3\sqrt{}\times3\sqrt{})$ R30° reconstruction. Superconducting proximity from an s-wave superconductor (Nb alloy), and magnetic proximity from a magnetic insulator (EuO), together create a topological superconductor channel on the system. (b) Electrons confined on the metal surface experience strong Rashba spin-orbit coupling and form two parabolas crossing at k=0. The gap of $2V_z$ comes from the exchange fields from EuO, and the chemical potential μ of the system needs to be tuned to the mid-gap. (c) Potential verification of the surface states through tunnelling spectrum, where one should see a dip corresponding to the opening of the exchange gap and a van Hove singularity corresponding to the very top of the Rashba bands.

The construction of the p-wave-superconductor-like electron systems makes the search for Majorana fermions much more tangible for experimentalists. It is now possible to revisit Kitaev's initial concept of 1D wires hosting Majorana end modes [36], which has been subsequently extended onto semiconductor nanowire platforms coupled with strong superconducting proximity effect [37, 38, 39]. The two ends of a 1D wire form a protected pair of Majorana fermions. This is analogous to a vortex on a 2D topological superconductor, which is also a boundary of the 2D platform but a physically moveable entity. To create a 1D p-wave superconductor wire, one can start with a 1D semiconductor wire and drive it into superconductivity with superconducting proximity effect. The regions surrounding the nanowire are just topologically trivial superconductors/insulators, leaving the enclosed topological superconductor nanowire intact. One can also start with the constructed 2D p-wave superconductor platforms as described earlier, and use selected electric and magnetic manipulation to suppress or flip the polarity of the topological superconductivity in order to define the desired 1D nanowire region with a distinct topology. Specifically, magnetic insulators can exert very strong interfacial exchange fields onto the desired 2D platforms, and can therefore be used as a local control to open and close the Zeeman gap of the system without much adverse influence on the superconductor system. Electric field manipulation from gate electrodes allows for local and active tuning of chemical potentials in the system. The topologically nontrivial superconductor state is enabled when the chemical potential is tuned into the mid gap, and will be destroyed when it is too far off. Therefore, we see that we can conveniently perform the fusion and braiding operations on the Majorana end modes with controlled gating on the 1D wire networks [39]. Similar manipulations can also be achieved on 2D platforms by controllably moving the vortices. Quantum information is stored non-locally on these non-Abelian systems and therefore is largely immune to any local perturbations, as long as the topology of the above systems remains protected by the superconducting and exchange gaps. These properties ensure that topological quantum computing on these solid-state platforms is inherently error-tolerant. We will not discuss in this article the details of braiding operations for topological quantum information processing, except to point out that quantum braids cover all the necessary quantum gates required for quantum computing. One can refer to the original article by Kitaev et al. [17] for an accurate account for topological quantum computation with non-Abelian anyons.

There have been a number of reports about successfully observing the signatures of Majorana fermions along the route of the proposed proximity-driven platforms [29, 30, 31, 40, 41, 42]. Semiconductors offer the most opportunities so far because of their maturity in technology. The first convincing report [29] was based on InSb nanowires with strong spin-orbit coupling and a very large g-factor ($g \approx 50$), and the device layout is shown in Fig.6. The nanowire portion under the superconductor is the region of interest, where the superconducting proximity effect drives this nanowire region into superconducting. Note that the superconductor only covers half of the wire in order that the electric fields coming from the buried gate electrodes are not completely screened. This practice is very important to ensure that the chemical potential of this portion of wire can be tuned into the middle of the Zeeman-split gap with controlled gating. The semiconductor nanowire with strong Rashba spin-orbit coupling can be treated as a quasi-1D electron system, and its dispersion relation is essentially the same as that illustrated in Fig.3's right-hand panel, with two Rashba-spin-split parabolas in zero magnetic fields. In the nanowire geometry, one can find that when the applied magnetic field is along the wire direction, a Zeeman energy gap opens up at the Dirac point of the dispersion curve. The above actions ensure that Majorana bound states will emerge at the ends of the wire when the topological superconductor phase disappears and the normal gapped semiconductor/superconductor phase emerges. A quantum mechanical tunnelling measurement on this Majorana mode confirmed that it is bound strictly at zero energy throughout the field range in which it is supported. It emerges when the field applied along the wire opens up the Zeeman gap, and disappears when the field becomes too strong and destroys the topological superconductor phase. The behaviour is distinctly different from other phenomena such as Andreev reflection and Kondo effect, and is therefore widely accepted as the first observation of Majorana bound states in solid-state systems. The bound state vanishes under elevated temperatures, because thermal excitations start to overcome the gap protections. Strictly speaking, a semiconductor nanowire is not exactly a 1D electron system, and there are multiple transverse subbands due to the finite width. The above 1D wire theory was expanded to the more general multichannel situations [34, 43], and the protected Majorana modes exist in the quasi-1D wires as long as there are only an odd number of bands crossing the Fermi level and the wire is not substantially wider than the superconducting coherence length. These conditions are both satisfied in this experiment. The magnetic field was applied along the wire direction in this experiment, and a magnetic field perpendicular to the device plane (out-of-the-page) is no longer a suitable choice. The circular geometry of the nanowires leads to a circular distribution of the Rashba electric fields, and for some parts of the wire surface the applied magnetic field would be aligned parallel to the wire surface and perpendicular to the wire direction. Magnetic field in such a configuration is only able to shift the Dirac point a little bit in energy and momentum, but is not able to open up the desired exchange gap. Therefore, some parts of the wire would remain gapless if a perpendicular-to-the-device-plane magnetic field is chosen.

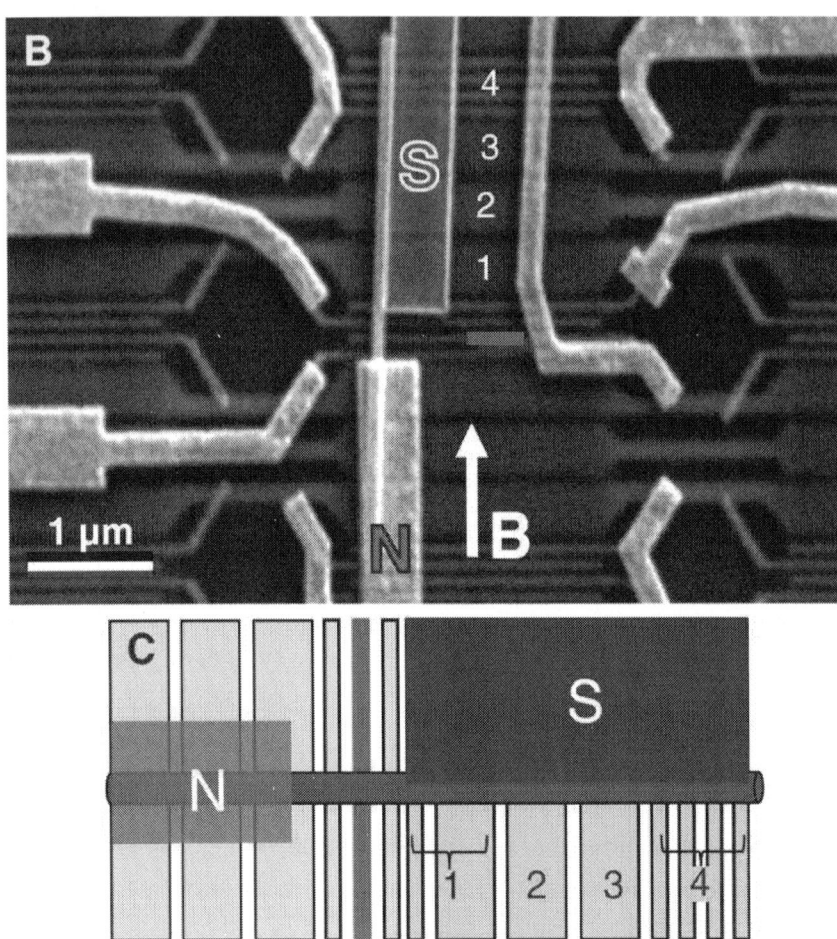

Figure 6. Device layout for the experimental evidence of Majorana fermion in a solid-state system [29]. The electrodes seen on the background, numbered 1, 2, 3, 4, etc., are the gate electrodes, separated from the nanowire and top electrodes by Si_3N_4 dielectrics. The nanowire is in contact with a normal metal (N) on one end, and a superconductor (S) on the other. The gate electrode labelled in dark green is the one used to create a tunnel barrier between the two sides of the nanowire. This effectively forms a S-NW-N tunnel junction for spectroscopy measurement. The nanowire region covered under the superconductor would have proximity-induced superconductivity in it, and a Majorana bound state is therefore trapped at the very edge of the superconductor region when the topology changes. It shows up as a zero-bias peak on the tunnel spectroscopy when a suitable magnetic field is applied along the wire direction.

INVERSE PROXIMITY ON SUPERCONDUCTORS — THE EXAMPLE OF SUPERCONDUCTING SPIN VALVES

As we have described above, the superconducting proximity effect describes the superconductor's wavefunction propagating into non-superconducting materials and inducing superconducting paring in there. Because of the leak-out of paired electrons, and the influx of unpaired electrons, the proximity effect would naturally generate a counter effect that weakens the superconductor itself. This is known as the inverse proximity effect. We will only focus our attention on a specific aspect of it — inverse proximity from ferromagnetic materials. The presence of spin information in the ferromagnets allows for spintronic manipulation on the superconductor system. For a detailed review on superconducting proximity with ferromagnets, please see the thorough review articles by Buzdin and Bergeret et al. [44, 45]. Here we proceed to demonstrate the inverse proximity effect from ferromagnets by illustrating with superconducting spin valve devices that can essentially turn the superconductivity On and Off through spin manipulations.

The propagation of superconductor wavefunction in a ferromagnetic metal is qualitatively different from that in a normal metal. This originates from the mutual incompatibility between a BCS superconductor, in which opposite spins pair up into Cooper pairs, and a ferromagnet, in which spins prefer to align parallel to each other to minimize the exchange energy. The presence of magnetic species in or near a superconductor material could be a drag if one wants to maximize the performance of the superconductor, because magnetic interactions break time reversal symmetry and become a strong source of Cooper pair breaking. We can, however, achieve desired controllability on the superconductor devices through manipulating the proximity-coupled magnetic systems. Fig.7 illustrates the propagation of superconducting wavefunction into a ferromagnetic metal. We note that instead of a simple exponential decay as in the normal metal/superconductor situation (Fig.1), the superconducting order parameter picks up a fast oscillatory behaviour. This can be qualitatively understood by considering the two electrons of a Cooper pair with opposite spins sitting inside an exchange field. One of the electrons, with its spin aligned along the exchange field direction, has its kinetic energy lowered by δE, while the opposite electron sees an increase in its kinetic energy by δE in the exchange field. Therefore, they start to propagate with different wavevector k, and the Cooper pair as a whole no longer has zero net momentum as it used to when it was inside the superconductor. For an electron near the Fermi level of a parabolic band, the kinetic energy is quadratic to the wavevector k and an increase (or decrease) in the kinetic energy by δE leads to a change in the wavevector by $\delta E/\hbar v_F$, where v_F is the Fermi velocity. The Cooper pair now gains a centre-of-mass momentum of $2\delta E/v_F$, and therefore the order parameter of the Cooper pairs will oscillate in space with a period of $hvF/2\delta E$. This roughly describes the system in the clean limit. For the dirty limit of a diffusive system, the Cooper pair motion can be described well with the Usadel equation [46]. Under the assumption that the exchange energy is much larger than the pairing energy, which is true for most ferromagnets relative to superconductors, the pair wave function can be solved and has a form of $\Delta \exp\left(-\frac{x}{\xi_f}\right)\cos\left(\frac{x}{\xi_f}\right)$ [44], i.e., an oscillatory function with exponential damping. In this special situation, the decay length and the oscillation period coincide (but they do not have to in more general situations). Here, $\xi f = \sqrt{D_f/\delta E}$. $D_f = \frac{1}{3}v_F l$ is the diffusion coefficient and l is the electrons' mean-free-path in the ferromagnet. The oscillatory behaviour of the pair wave function adds an oscillatory dependence to the structure's overall critical temperature [47] when the ferromagnet layer thickness is comparable or longer than the pair oscillating period ξ_F. This has been verified in various superconductor-ferromagnet bilayer structures [48, 49, 50, 51]. One can readily see that when the ferromagnet layer thickness is equal to $\pi\xi_F$, the wavefunction has a phase shift of π across the ferromagnet layer, and the superconductivity is therefore weakened. As the ferromagnet increases in thickness, the phase oscillates back and forth and the system shows an oscillatory, non-monotonic behaviour in the critical temperature. The parameter ξ_F can be approximately read out from the oscillation period of T_C. The oscillatory behaviour of the pair wavefunction shows up not only in the phases, but also in the electron density of states inside the ferromagnets. At certain distance away from the interface, the density of states can even become larger than the normal value in the absence of superconducting proximity [52].

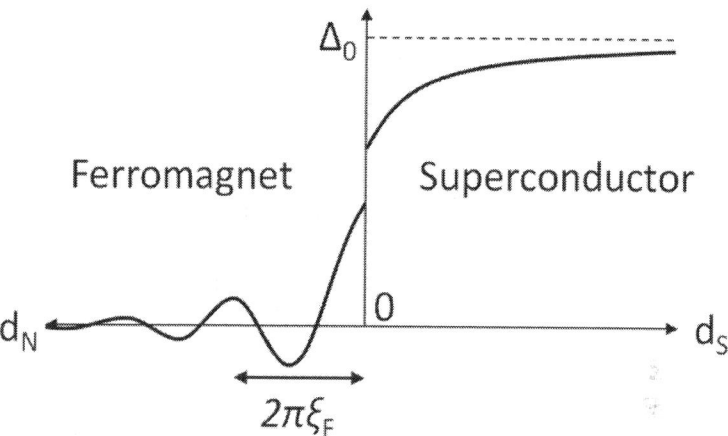

Figure 7. Evolution of the superconductor pairing potential near a S/F interface. The pairing potential inside the ferromagnet shows a strong oscillatory and damping behaviour due to the large internal exchange field. ξ_F characterizes the superconducting correlation oscillation length as well as the correlation decay length within a strong ferromagnet, where the exchange energy δE far exceeds the pairing potential Δ.

Though it is not immediately clear, the above discussions already imply that superconductivity within the superconductor itself will also be significantly weakened in the presence of a strong ferromagnet. A critical difference of the proximity with a strong ferromagnet, compared to that with a normal metal, is the extremely short decay length of the pairing potential. By connecting the superconducting wavefunction across the interface, one can already see that the steep slope of the pairing potential on the ferromagnet side influences that on the superconductor side and pulls the whole pairing potential profile downward. The exact boundary condition matching is subject to the interfacial transparency and the respective diffusion coefficients. A stronger ferromagnet creates stronger exchange interaction and shorter ξ_F: therefore, stronger suppression to the superconductivity. Now we can consider constructing a sandwich of ferromagnet/superconductor/ferromagnet: i.e., a superconducting spin valve structure [53, 54]. Although the two ferromagnet layers will both strongly suppress superconductivity in the sandwiched middle layer, we can try to cancel out their influence through spin manipulation. Recall that when one of the ferromagnet layers is placed on top of the superconductor, strong suppression of superconductivity is present. If we place the second ferromagnet layer with the same spin polarity, the inverse proximity effects coming from both ferromagnets strengthen each other and suppress superconductivity even more. This spin-parallel configuration ensures that the wavefunction over the whole structure is symmetric with respect to the film centre (mid-plane of the superconductor). We can conceptually fold the wavefunction in the middle and it becomes identical to that of a bilayer situation. Because the equivalent thickness of the superconductor layer is only half of its actual thickness, the layer suffers even stronger suppression from the inverse proximity effect. If we instead place the second ferromagnet with an opposite spin polarity relative to the first one, i.e., configuring the system into a spin-antiparallel state, the proximity effects from the two ferromagnets differ precisely by a phase of π and will largely cancel each other out. Superconductivity is therefore restored in this configuration.

We now use an epitaxial pristine structure of Fe/V/Fe as an example to illustrate the above mentioned superconducting spin valve effect [51]. The structure was deposited with molecular beam epitaxy (MBE) onto MgO-buffered Si(100) substrates. X-ray diffraction verified the epitaxy of the films showing clear four-fold symmetry in the off-axis diffraction patterns. The quality of the films and interfaces can be verified with the critical temperature T_C and upper critical field H_{C2} measurements on V single layer films and Fe/V bilayer films. Fig.8 (a) shows that T_C varies linearly with respect to the inverse film thickness. Such dependence indicates clearly that there exists a superconducting "dead" layer on the film surface due to the lowering of electron density and weakening of phonon coupling there [55]. This surface layer is superconductingly inactive, and amounts to about 1 atomic layer (0.18 nm) on the pure V films (interfaced with MgO on both sides), confirming the superior quality of these films. On the other hand, the series of bilayer samples of V with proximity to 6 nm Fe show a significantly steeper slope, corresponding to a much thicker inactive layer of 13.5 nm. For V of the same thickness, presence of Fe pulls T_C much lower than MgO does in pure V films. This is a clear manifestation of the above-described inverse superconducting proximity effect: the presence of a ferromagnet strongly suppresses superconductivity in the adjacent superconductor. Fig.8 (b) shows the upper critical fields of the set of films with 30 nm V. The addition of proximity from Fe again significantly suppresses superconductivity. The linear dependence of H_{C2} square with respect to temperature is an indication of two-dimensional superconductivity behaviour, which follows HC2(T)=HC2(0)(1−TTC)1/2 [56]. From the determined slope of the plot, we can identify the superconducting coherence length ξ_S, being about 8.2 nm for the 30 nm V films. It is interesting to note that the slope does not change when the proximity effect from Fe kicks in, indicating that no change happens to the Cooper pair correlations. This is quite expected because the proximity effect and inverse proximity effect do not modify the pairing mechanisms within the superconductor, and the addition of a few nm metals has little influence on the electrons' mean-free-paths either. Fig.8 (c) shows the most important aspects of the inverse superconducting proximity effect: the quick oscillation and damping of T_C with thickness. As we have described earlier, T_C oscillates with the ferromagnetic layer thickness because of the Cooper pairs' phase oscillation inside an exchange field. We can roughly estimate the superconducting correlation length inside the ferromagnet ξ_F to be about 1 nm. We can also clearly see that the superconductivity suppression from inverse proximity effect is quite dramatic: with a few monolayers of Fe overlayer, the critical temperature already drops from the original value of over 5K to only slightly above 3K. The oscillation quickly damps out after several ξ_F, when the Cooper pair wavefunction is almost fully suppressed. After the first few nm, further increase of the ferromagnetic layer thickness does not continue to influence the overall wavefunction. On the other hand, we notice that T_C does not really drop to zero before recovering, as we have depicted before. This actually is also a consequence of the fast damping behaviour which significantly limits the transparency of the Cooper pairs into the ferromagnet. Beyond the first few nm, the remaining portion of the ferromagnet has no significant further contribution to the T_C suppression. In addition, because the measurement of T_C is over the whole structure, the region of the superconductor far away from the interface essentially retains the original superconductivity strength, and renders the whole structure superconducting under conventional DC measurements. Detailed calculations on T_C of such ferromagnet/superconductor bilayer structures confirm that the oscillation indeed quickly reaches a saturation value after the ferromagnet thickness increases beyond the first few oscillations [57].

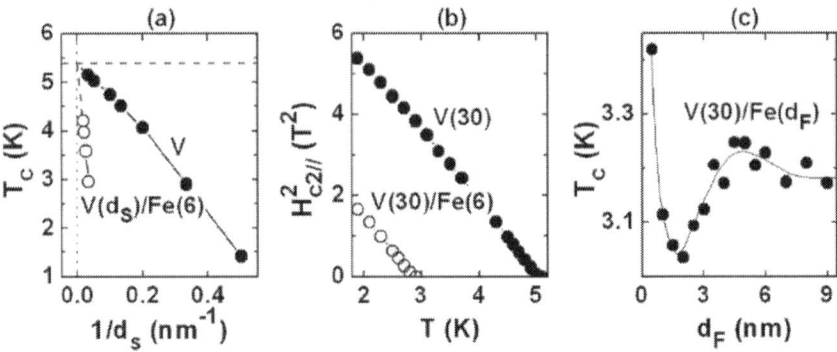

Figure 8. Critical temperature T_C and upper critical field H_{C2} for epitaxial V and Fe/V films [51]. (a) T_C as a function of inverse V layer thickness; (b) H_{C2} square as a function of temperature; (c) T_C damping and oscillation with respect to the Fe layer thickness, the solid line is provided as a guide for the eye. ξ_F can be estimated to be about 1 nm.

Next, we consider the magneto resistance of these superconducting spin valves. The structure is fully epitaxial, with high interfacial quality. In order to magnetically separate the two Fe layers, a CoO anti ferromagnetic layer was introduced above the top Fe layer, which exerts exchange bias [58] on the top Fe layer and essentially pins the magnetization of this layer to a chosen direction. When the applied magnetic field only sweeps in a range not exceeding the exchange bias pinning strength, one can only observe magnetic switching of the "free" layer, which is the bottom Fe layer in this structure. Fig.9 (a) shows the magneto resistance results across temperatures close to T_C. When the magnetic field varies, clear resistance high and low states are observed, indicating that the structure toggles between the non-superconducting and superconducting states. In the spin-parallel configuration, the inverse proximity effects from both ferromagnet layers strengthen each other and strongly suppress superconductivity of the sandwiched V layer, leading to the high-resistance, non-superconducting state; in the spin-anti parallel configuration, the inverse proximity effects largely cancel each other out and the system returns to the low resistance, superconducting state. For certain temperatures the ratio between the normal resistance and the superconducting resistance is essentially infinite, making the magneto resistance ratio essentially infinite. The effect is only pronounced in a very narrow temperature window close to T_C, because this is when the superconductivity is already quite weakened by thermal energy, and slight perturbations from the environment (such as from the inverse proximity effects) can readily drive the system in and out of its superconducting state.

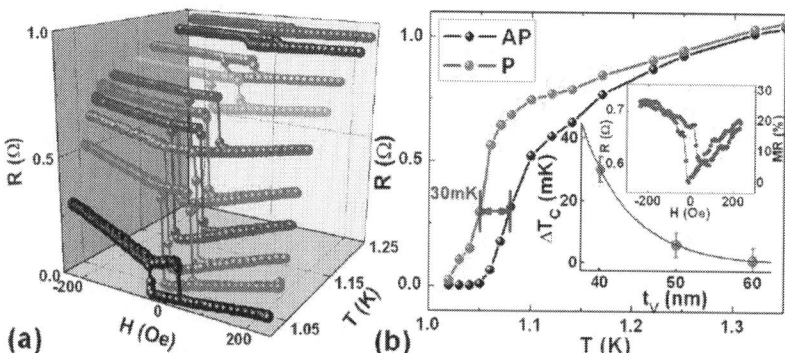

Figure 9. Measured superconducting spin valve effect in the epitaxial Fe/V/Fe structure [51]. (a) Device resistance as a function of applied magnetic field H and temperature T. A large magnetoresistance is associated with the switching of the free Fe layer in the temperature range close to T_C. (b) Resistance as a function of temperature for the structure in its spin-parallel (P) and spin-antiparallel (AP) configurations. There is a clear offset of 30mK between their corresponding T_C's. Inset: $T_C$offset as a function of the V layer thickness, and an example of a magnetoresistance curve for the sample with 50 nm V.

Fig.9 (b) shows T_C determined for the two spin configurations. Clearly, spin-parallel state has weaker superconductivity and lower T_C, while spin-antiparallel state has stronger superconductivity and higher T_C. In the temperature range between these two T_C, a large resistance difference exists between the two states leading to the large magnetoresistance described above. The superconducting spin valve effect relies on the fact that the two ferromagnet layers are effectively coupled to each other through the superconductor. Therefore, the effect decreases as the superconductor layer thickness increases, and essentially drops to zero when the superconductor layer is much thicker than the superconducting coherence length ξ_S of about 8 nm in this system. On the other hand, if the superconductor layer is too thin, its superconductivity would be fully suppressed by the Fe layers and the superconducting spin valve effect is also no longer achievable. The T_C shift as seen in Fig.9 (b) is clear indication of the operation temperature range of this effect, which is, however, limited to only tens of mK across many different systems [59, 60, 61, 62,51]. Improving interfacial transparency is the key for further improving the performance of these devices. Fig.10 shows the calculated variation of the P and AP T_C's with the change of the interfacial transparency [53]. The curves with the highest transparency (solid lines, $T_F=25$) clearly show a region with very large T_C shift. For example, around $d_F/\xi_F=0.4$, T_C of the AP state is around half the bulk value while that of the P state is fully suppressed. For samples of this structure, infinite magnetoresistance would show up for any temperatures below half the bulk T_C, making it a practically usable device for turning the superconductivity On and Off. As a comparison, for the curves corresponding to the lowest interfacial transparency, $T_F = 1$, the $T_C$difference between P and AP states is very small throughout all the d_F choices. This is easy to understand because poor interface transparency breaks up the correlation between the superconductor and the ferromagnets: therefore, the correlation between the two ferromagnets also vanishes. The two ferromagnets do not feel much influence from each other and the P and AP configurations make little difference. Although a weak superconducting spin valve effect is still present, the resultant magnetoresistance is very small and there is no truly On and Off tuning of superconductivity, making it less useful in practice.

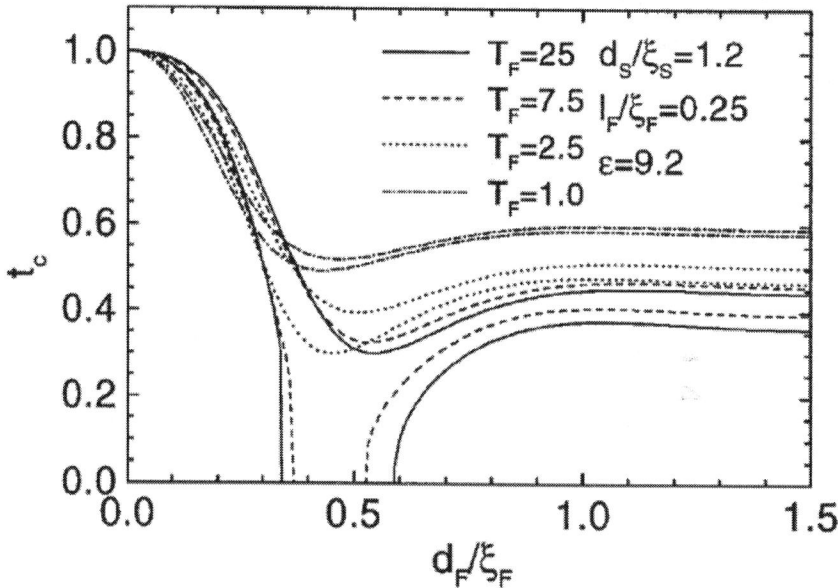

Figure 10. Expected T_C shift in superconducting spin valves when the interface transparency is varied [53]. Here d_S and d_F are the superconductor and ferromagnet layer thicknesses, l_F is the electron mean-free-path in the ferromagnet, ξ_S and ξ_F are the superconducting correlation lengths in the superconductor and ferromagnet, and ε is a parameter depending on the properties of both the superconductor and the ferromagnet, getting smaller with stronger ferromagnets. Two curves are shown for each choice of interfacial transparency, with the upper curve corresponding to the AP state and the lower one to the P state.

We next examine a special type of superconducting spin valve, where the ferromagnet layers are not metallic but insulating [63]. The above described inverse superconducting proximity mechanism does not apply because the Cooper pair wavefunction cannot penetrate into the insulators. The inverse proximity happens due to the interfacial exchange interactions from the localized moments of the magnetic insulator, which exert a large effective Zeeman field onto the superconductor and can also suppress the superconductivity [32]. The interaction between the magnetic insulator and the carriers in the superconductor is through indirect exchange interaction, where the free electrons communicate with multiple localized magnetic moments of the magnetic insulator, and become spin-polarized in the process. This behaves as an effective Zeeman field on the electron system and favours one type of spins over the other, therefore counteracts the tendency of Cooper pairing and suppresses superconductivity. Now that the magnetism-superconductivity interaction is established, it is again possible to use spin manipulations to control superconductivity in a sandwich structure [64]. The magnetic insulator material used in this study was EuS, which has been shown to generate exchange fields as large as a few Tesla on thin superconducting Al films [65, 66]. Under this effective Zeeman field on the conduction electrons, Cooper paring in the superconductor becomes less stable and superconductivity is thus weakened accordingly. A second magnetic insulator layer, when configured with its magnetic orientation opposite to the first one, will serve to cancel the exchange fields from the first one, provided that the electrons' mean-free-paths are long enough relative to the superconductor layer thickness such that the Zeeman fields from one side can propagate across the whole superconductor layer and influence the other side. As a result, we again recover a spin valve performance: when the two magnetic insulators are aligned parallel to each other, their exchange fields stack up and destroy the superconductivity in the system; when they are aligned antiparallel to each other, their effects cancel and the system goes back to the superconducting state. Experimentally, toggling the superconductivity On and Off leads to a very large magnetoresistance response, as shown in Fig.11. Here the system is in the spin-parallel state (finite resistance, non-superconducting state) when the external field is large and both EuS layers are saturated in the same direction; and in the spin-antiparallel state (zero resistance, superconducting state) when the external field is tuned between the magnetic switching fields of the two EuS layers, such that one of the layers flips while the other does not. Similar to the superconducting spin valves with magnetic metals, this effect only shows up in a very narrow temperature window close to T_C, when the superconductivity is already very weak to begin with. Though this effect appears very similar to the previously described superconducting spin valve effect, there is one critical difference: in this structure, there are negligible superconducting proximity and inverse proximity effects because of the insulating nature of the ferromagnets. The observed magnetic tuning of superconductivity is in fact a result of the propagation of magnetic exchange interaction into the superconductor — the magnetic proximity effect, which is especially pronounced with magnetic insulators [32]. As a result, the system is described with the Cooper pairs feeling the average exchange fields *inside the superconductor* rather than with Cooper pairs feeling the exchange fields *inside the ferromagnets*.

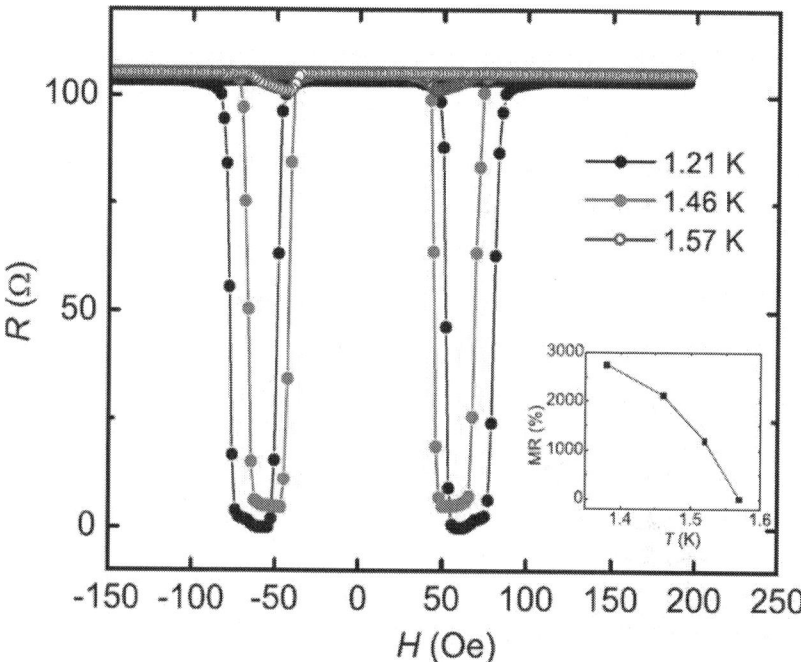

Figure 11. Superconducting spin valve effect in a magnetic insulator/ superconductor / magnetic insulator sandwich structure (EuS/Al/EuS) [64]. Inset shows the variation of magnetoresistance ratio over the temperatures close to T_C.

SUMMARY

In summary, we have used a few very specific examples (the creation of Majorana fermions on 2D surface states and the superconducting spin valves with ferromagnets) to illustrate the application of superconducting proximity and inverse proximity effects in spintronics. These examples can be viewed as passive devices of superconducting proximity/inverse proximity. One can also use active pumping of Cooper pairs to drive supercurrents into non-superconducting materials, which is yet another manifestation of the superconducting proximity effect and can induce superconducting Josephson coupling on many spintronics platforms, such as graphene and topological insulators. The Josephson effect will be covered in other chapters of this book and we will not discuss it in this chapter. Overall, we see that the superconducting proximity and inverse proximity effects are convenient approaches to couple superconductivity with many other types of spin systems, and allow us to create hybrid devices that can benefit from these very distinct spin states: superconductivity, magnetism, and topological quantum spin Hall state. Such manipulation of superconductivity offers important new routes for information storage and processing, and rapid advances are expected to happen in these directions taking information processing to the quantum level. We, as well as many other groups in the world, are actively working toward such goals.

ACKNOWLEDGEMENTS

The author wish to thank the Natural Sciences and Engineering Research Council of Canada (NSERC) for support on this work.

REFERENCES

1. Meissner H, Superconductivity of Contacts with Interposed Barriers, Phys. Rev. 1960; 117: 672.
2. de Gennes P G, Boundary Effects in Superconductors, Rev. Mod. Phys. 1964; 36: 225.
3. Takayanagi H, Kawakami T, Superconducting Proximity Effect in the Native Inversion Layer on InAs, Phys. Rev. Lett. 1985; 54: 2449.
4. Nitta J, Akazaki T, Takayanagi H, Arai K, Transport Properties in an InAs-inserted-channel In0.52Al0.48As / In0.53Ga0.47As Heterostructure Coupled Superconducting Junction, Phys. Rev. B 1992; 46: 14286(R).
5. Nguyen C, Kroemer H, Hu E L, Anomalous Andreev Conductance in InAs-AlSb Quantum Well Structures with Nb Electrodes, Phys. Rev. Lett. 1992; 69: 2847.

6. Zhang D, Wang J, DaSilva A M, Lee J S, Gutierrez H R, Chan M H W, Jain J, Samarth N, Superconducting Proximity Effect and Possible Evidence for Pearl Vortices in a Candidate Topological Insulator, Phys. Rev. B 2011; 84: 165120.

7. Williams J R, Bestwick A J, Gallagher P, Hong S S, Cui Y, Bleich A S, Analytis J G, Fisher I R, Goldhaber-Gordon D, Unconventional Josephson Effect in Hybrid Superconductor-Topological Insulator Devices, Phys. Rev. Lett. 2012; 109: 056803.

8. Qu F, Yang F, Shen J, Ding Y, Chen J, Ji Z, Liu G, Fan J, Jing X, Yang C, Lua L, Strong Superconducting Proximity Effect in Pb-Bi2Te3 Hybrid Structures, Sci Rep. 2012; 2: 339.

9. Heersche H B, Jarillo-Herrero P, Oostinga J B, Vandersypen L M K, Morpurgo A F, Bipolar Supercurrent in Graphene, Nature 2007; 446: 56–59.

10. Morpurgo A F, Kong J, Marcus C M, Dai H, Gate-Controlled Superconducting Proximity Effect in Carbon Nanotubes, Science 1999; 286 (5438): 263–265.

11. Jarillo-Herrero P, van Dam J A, Kouwenhoven L P, Quantum Supercurrent Transistors in Carbon Nanotubes, Nature 2006; 439: 953–956.

12. Winkelmann C B, Roch N, Wernsdorfer W, Bouchiat V, Balestro F, Superconductivity in a Single-C60 Transistor, Nature Physics 2009; 5: 876–879.

13. van Dam J A, Nazarov Y V, Bakkers E P A M, De Franceschi S, Kouwenhoven L P, Supercurrent Reversal in Quantum Dots, Nature 2006; 442: 667–670.

14. Kasumov A Y, Kociak M, Guéron S, Reulet B, Volkov V T, Klinov D V, Bouchiat H, Proximity-Induced Superconductivity in DNA, Science 2001; 291 (5502): 280–282.

15. Peña V, Sefrioui Z, Arias D, Leon C, Santamaria J, Varela M, Pennycook S J, Martinez J L, Coupling of Superconductors Through a Half-Metallic Ferromagnet: Evidence for a Long-Range Proximity Effect, Phys. Rev. B 2004; 69: 224502.

16. Keizer R S, Goennenwein S T B, Klapwijk T M, Miao G X, Xiao G, Gupta A, A Spin Triplet Supercurrent Through the Half-Metallic Ferromagnet CrO2, Nature 2006; 439(7078): 825.

17. Kitaev A Yu, Fault-Tolerant Quantum Computation by Anyons, Annals. Phys. 2003; 303(1): 2–30.

18. Read N, Green D, Paired States of Fermions in Two Dimensions with Breaking of Parity and Time-Reversal Symmetries and the Fractional Quantum Hall Effect, Phys. Rev. B 2000; 61: 10267.

19. Ivanov D A, Non-Abelian Statistics of Half-Quantum Vortices in p-Wave Superconductors, Phys. Rev. Lett. 2001; 86: 268.

20. Das Sarma S, Nayak C, Tewari S, Proposal to Stabilize and Detect Half-Quantum Vortices in Strontium Ruthenate Thin Films: Non-Abelian Braiding Statistics of Vortices in a px+ipy Superconductor, Phys. Rev. B 2006; 73: 220502(R).

21. Mackenzie A P, Maeno Y, The Superconductivity of Sr2RuO4 and the Physics of Spin-Triplet Pairing, Rev. Mod. Phys. 2003; 75: 657.

22. Fu L, Kane C L, Superconducting Proximity Effect and Majorana Fermions at the Surface of a Topological Insulator, Phys. Rev. Lett. 2008; 10 (9): 096407.

23. Sau J D, Lutchyn R M, Tewari S, Das Sarma S, Generic New Platform for Topological Quantum Computation Using Semiconductor Heterostructures, Phys. Rev. Lett. 2010; 104(4): 040502.

24. Hasan M Z, Kane C L, Colloquium: Topological Insulators, Rev. Mod. Phys. 2010; 82: 3045.

25. Fu L, Kane C L, Mele E J, Topological Insulators in Three Dimensions, Phys. Rev. Lett. 2007; 98: 106803.

26. He L, Kou X, Wang K L, Review of 3D Topological Insulator Thin-Film Growth by Molecular Beam Epitaxy and Potential Applications, phys. stat. sol. - Rap. Res. Lett. 2013; 7: 50–63.

27. Sasaki S, Kriener M, Segawa K, Yada K, Tanaka Y, Sato M, Ando Y, Topological Superconductivity in CuxBi2Se3

28. Wang M X, Liu C, Xu J P, Yang F, Miao L, Yao M Y, Gao C L, Shen C, Ma X, Chen X, Xu Z A, Liu Y, Zhang S H, Qian D, Jia J F, Xue Q K, The Coexistence of Superconductivity and Topological Order in the Bi2Se3 Thin Films, Science 6 April 2012; 336: 52–55.

29. Mourik V, Zuo K, Frolov S M, Plissard S R, Bakkers E P A M, Kouwenhoven L P, Signatures of Majorana Fermions in Hybrid Superconductor-Semiconductor Nanowire Devices, Science 2012; 336 (6084): 1003–1007.

30. Deng M T, Yu C L, Huang G Y, Larsson M, Caroff P, Xu H Q, Anomalous Zero-Bias Conductance Peak in a Nb–InSb Nanowire–Nb Hybrid Device, Nano Lett. 2012; 12, 6414–6419.

31. Das A, Ronen Y, Most Y, Oreg Y, Heiblum M, Shtrikman H, Zero-bias Peaks and Splitting in an Al-InAs Nanowire Topological Superconductor as a Signature of Majorana Fermions, Nature Physics 2012; 8: 887–895.

32. Miao G X, Moodera J S, Spin Manipulation with Magnetic Semiconductor Barriers, Phys. Chem. Chem. Phys. 2014; DOI: 10.1039/c4cp04599h, in print.

33. Alicea J, Majorana Fermions in a Tunable Semiconductor Device, Phys. Rev. B 2010; 81: 125318.

34. Potter A C, Lee P A, Multichannel Generalization of Kitaev's Majorana End States and a Practical Route to Realize Them in Thin Films, Phys. Rev. Lett. 2010; 105: 227003.

35. Law K T, Lee P A, Ng T K, Majorana Fermion Induced Resonant Andreev Reflection, Phys. Rev. Lett. 2009; 103: 237001.

36. Kitaev A Y, Unpaired Majorana Fermions in Quantum Wires, Physics-Uspekhi 2001; 44: 131.

37. Lutchyn R M, Sau J D, Das Sarma S Majorana Fermions and a Topological Phase Transition in Semiconductor-Superconductor Heterostructures. Phys. Rev. Lett. 2010; 105 (7): 077001.

38. Oreg Y, Refael G, von Oppen F, Helical Liquids and Majorana Bound States in Quantum Wires, Phys. Rev. Lett. 2010; 105: 177002.

39. Alicea J, Oreg Y, Refael G, von Oppen F, Fisher M P A, Non-Abelian Statistics and Topological Quantum Information Processing in 1D Wire Networks, Nature Physics 2011; 7: 412–417.

40. Nadj-Perge S, Drozdov I K, Li J, Chen H, Jeon S, Seo J, MacDonald A H, Bernevig B A, Yazdani A, Observation of Majorana Fermions in Ferromagnetic Atomic Chains on a Superconductor, Science 2014; 346(6209): 602–607.

41. Stanescu T D, Tewari S, Majorana Fermions in SemiconductorNanowires: Fundamentals, Modeling, and Experiment, J. Phys.: Condens. Matter 2013; 25: 233201.

42. Beenakker C W J, Search for Majorana Fermions in Superconductors. Annu. Rev. Condens. Matter Phys. 2013; 4: 113.

43. Lutchyn R M, Stanescu T D, Das Sarma S, Search for Majorana Fermions in Multiband Semiconducting Nanowires, Phys. Rev. Lett. 2011; 106: 127001.

44. Buzdin A I, Proximity Effects in Superconductor-Ferromagnet Heterostructures, Rev. Mod. Phys. 2005; 77: 935–976.

45. Bergeret F S, Volkov A F, Efetov K B, Odd Triplet Superconductivity and Related Phenomena in Superconductor-Ferromagnet Structures, Rev. Mod. Phys. 2005; 77: 1321–1373.

46. Usadel K D, Generalized Diffusion Equation for Superconducting Alloys, Phys. Rev. Lett. 1970; 25: 507.

47. Buzdin A I, Kuprianov M Y, Transition Temperature of a Superconductor-Ferromagnet Superlattice, Pis'ma Zh. Eksp. Teor. Fiz. 1990; 52, 1089–1091.

48. Jiang J S, Davidović D, Reich D H, Chien C L, Oscillatory Superconducting Transition Temperature in Nb/Gd Multilayers, Phys. Rev. Lett. 1995; 74: 314.

49. Mercaldo L V, Attanasio C, Coccorese C, Maritato L, Prischepa S L, Salvato M, Superconducting-Critical-Temperature Oscillations in Nb/CuMn Multilayers, Phys. Rev. B 1996; 53: 14040.

50. Sidorenko A S, Zdravkov V I, Prepelitsa A A, Helbig C, Luo Y, Gsell S, Schreck M, Klimm S, Horn S, Tagirov L R, Tidecks R, Oscillations of the Critical Temperature in Superconducting Nb/Ni bilayers, Annalen der Physik 2003; 12: 37–50.

51. Miao G X, Ramos A V, Moodera J S, Infinite Magnetoresistance from the Spin Dependent Proximity Effect in Symmetry Driven bcc-Fe/V/Fe Heteroepitaxial Superconducting Spin Valves,Phys. Rev. Lett. 2008; 101: 137001.

52. Kontos T, Aprili M, Lesueur J, Grison X, Inhomogeneous Superconductivity Induced in a Ferromagnet by Proximity Effect, Phys. Rev. Lett. 2001; 86: 304.

53. Tagirov L R, Low-Field Superconducting Spin Switch Based on a Superconductor/Ferromagnet Multilayer, Phys. Rev. Lett. 1999; 83: 2058.

54. Buzdin A I, Vedyayev A V, Ryzhanova N V, Spin-Orientation-Dependent Superconductivity in F/S/F Structures, Europhys. Lett. 1999; 48: 686.

55. Wolf S A, Kennedy J J, Nisenoff M, Properties of Superconducting rf Sputtered Ultrathin Films of Nb,J. Vac. Sci. Technol. 1976; 13: 145.

56. Tinkham M, Introduction to Superconductivity, McGraw-Hill, New York, 1978.

57. Demler E A, Arnold G B, Beasley M R, Superconducting Proximity Effects in Magnetic Metals, Phys. Rev. B 1997; 55: 15174.

58. Berkowitz A E, Takano K, Exchange Anisotropy — A Review, J. Magn. Magn. Mater. 1999; 200: 552–570.

59. Gu J Y, You C Y, Jiang J S, Pearson J, Bazaliy Ya B, Bader S D, Magnetization-Orientation Dependence of the Superconducting Transition Temperature in the Ferromagnet-Superconductor-Ferromagnet System: CuNi/Nb/CuNi, Phys. Rev. Lett. 2002; 89: 267001.

60. Potenza A, Marrows C H, Superconductor-Ferromagnet CuNi/Nb/CuNi Trilayers as Superconducting Spin-Valve Core Structures, Phys. Rev. B 2005; 71: 180503(R).

61. Moraru I C, Pratt W P, Birge N O, Magnetization-Dependent Tc Shift in Ferromagnet/Superconductor/Ferromagnet Trilayers with a Strong Ferromagnet, Phys. Rev. Lett. 2006; 96: 037004.

62. Nowak G, Zabel H, Westerholt K, Garifullin I, Marcellini M, Liebig A, Hjörvarsson B, Superconducting Spin Valves Based on Epitaxial Fe/V Superlattices, Phys. Rev. B 2008; 78: 134520.

63. de Gennes P G, Coupling Between Ferromagnets Through a Superconducting Layer, Phys. Lett. 1966; 23: 10–11.

64. Li B, Roschewsky N, Assaf B A, Eich M, Epstein-Martin M, Heiman D, Munzenberg M, Moodera J S, Superconducting Spin Switch with Infinite Magnetoresistance Induced by an Internal Exchange Field, Phys. Rev. Lett. 2013; 110: 097001.

65. Moodera J S, Hao X, Gibson G A, Meservey R, Electron-Spin Polarization in Tunnel Junctions in Zero Applied Field with Ferromagnetic EuS Barriers, Phys. Rev. Lett. 1988; 61: 637.

66. Hao X, Moodera J S, Meservey R, Spin-Filter Effect of Ferromagnetic Europium Sulfide Tunnel Barriers, Phys. Rev. B 1990; 42: 8235.

CITATION

Guoxing Miao (2015). Spintronics Driven by Superconducting Proximity Effect, Superconductors - New Developments, Dr. Alexander Gabovich (Ed.), ISBN: 978-953-51-2133-6, InTech, DOI: 10.5772/59942.

CHAPTER 10

Europium Doped DI-Calcium Magnesium DI-Silicate Orange–Red Emitting Phosphor by Solid State Reaction Method

Ishwar Prasad Sahu, D.P. Bisen and Nameeta Brahme

School of Studies in Physics & Astrophysics, Pt. Ravishankar Shukla University, Raipur, C.G., 492010, India

ABSTRACT

A new orange–red europium doped di-calcium magnesium di-silicate ($Ca_2MgSi_2O_7$:Eu^{3+}) phosphor was prepared by the traditional high temperature solid state reaction method. The prepared $Ca_2MgSi_2O_7$:Eu^{3+} phosphor was characterized by X-ray diffractometer (XRD), transmission electron microscopy (TEM), field emission scanning electron microscopy (FESEM) with energy dispersive x-ray spectroscopy (EDX), fourier transform infrared spectra (FTIR), photoluminescence (PL) and decay characteristics. The phase structure of sintered phosphor was akermanite type structure which belongs to the tetragonal crystallography with space group $P\bar{4}2_1m$, this structure is a member of the melilite group and forms a layered compound. The chemical composition of the sintered $Ca_2MgSi_2O_7$:Eu^{3+} phosphor was confirmed by EDX spectra. The PL spectra indicate that $Ca_2MgSi_2O_7$:Eu^{3+} can be excited effectively by near ultraviolet (NUV) light and exhibit bright orange–red emission with excellent color stability. The fluorescence lifetime of $Ca_2MgSi_2O_7$:Eu^{3+} phosphor was found to be 28.47 ms. CIE color coordinates of $Ca_2MgSi_2O_7$:Eu^{3+} phosphor is suitable as orange-red light emitting phosphor with a CIE value of (X = 0.5554, Y = 0.4397). Therefore, it is considered to be a new promising orange–red emitting phosphor for white light emitting diode (LED) application.

INTRODUCTION

Luminescent materials containing rare earth ions are able to absorb energy in the UV-regions and emit visible light. Recently, these materials have drawn increasing interest due to their promising applications in white light emitting diodes, display devices, storage bioluminescence and fluorescence labels (Sahu et al., 2015b and Xu et al., 2014).

As a new solid state light source, the white light-emitting diodes (WLEDs) are considered to be the fourth generation general lighting devices that stands a real chance of replacing conventional lighting sources such as incandescent and fluorescent lamps due to its long lifetime, saving energy, reliability, safety and its environmental friendly characteristics (Jiao & Wang, 2012). Among the technological strategies of obtaining WLEDs, the phosphor converted (pc) emission method is the most common one, in which tricolor phosphors (red, green and blue) are pumped by UV InGaN chips or blue GaN chips and generate white light. At present, the commercially used green and red phosphors for NUV chips are ZnS: (Cu^+, Al^{3+}) and $Y_2O_2S:Eu^{3+}$, respectively. Unfortunately, both ZnS: (Cu^+, Al^{3+}) and $Y_2O_2S:Eu^{3+}$ show low chemical stability as they are sulfide based phosphors. Therefore, it is urgent to search for new green and red phosphors or an orange–red phosphor with high efficiency and excellent stability (Liu et al. 2014).

Generally, phosphors consist of activator and host, in order to obtain efficient red or orange–red emitting phosphor, host is another key factor (Wang, Lou, & Li, 2014). Eu^{3+} doped oxides were widely studied as efficient red emitting phosphors due to the abundant transitions from the excited $^5D_0$ level to the ^7F_J (J = 0, 1, 2, 3, 4) levels of the $4f^6$ configuration in the orange-red light area (Dong et al., 2014 and Gorller-Walrand et al., 1995). Mellite are a large group of compounds characterized by the general formula $M_2T^1T^2_2O_7$, (M = Sr, Ca, Ba; T^1 = Mn, Co, Cu, Mg, Zn; T_2 = Si, Ge), have been investigated widely as optical materials. Due to their tetragonal and non-centrosymmetric crystal structure, lanthanides or transition metals can be accepted easily as constituents or dopants by the melilites, allowing the synthesis of high-quality doped single crystals. Recently, di-calcium magnesium di-silicate ($Ca_2MgSi_2O_7$) phosphor has attracted great interest due to its special structure features, excellent physical and chemical stability. They have been studied widely with Eu^{2+} doping, which shows that a green emission and long persistent luminescence by co-doping with some other rare earth ions. A calcium silicate phosphor would be ideal from the manufacturing point of view, because both calcium and silica are abundant and are relatively inexpensive (Talwar et al. 2009).

In the past, $Ca_2MgSi_2O_7$ phosphor doped with Eu^{3+} has been prepared by pulsed laser deposition method. High quality $Ca_2MgSi_2O_7:Eu^{3+}$ films phosphors were deposited on Al_2O_3 (0 0 0 1) substrates. The crystallinity, surface roughness and photoluminescence of the thin film phosphors were strongly dependent on the deposition conditions, which is the drawback of pulsed laser deposition method (Yang, Moona, Choi, Jeong, & Kim, 2012). Solid state reaction techniques is a traditional phosphors synthesis techniques is widely used to prepare silicate phosphors because samples prepared using this method have good luminescence and very good morphology, which has advantage over the pulsed laser deposition technique (Sahu et al., 2015a and Shrivastava and Kaur, 2014).

In the present paper, we report the synthesis of europium doped di-calcium magnesium di-silicate ($Ca_2MgSi_2O_7:Eu^{3+}$) phosphor by high temperature solid state reaction method. The phase structure, crystallite size, particle size, surface morphology, elemental analysis, different stretching mode was analyzed by X-ray diffractometer (XRD), transmission electron microscopy (TEM), field emission scanning electron microscopy (FESEM) with energy dispersive X-ray spectroscopy (EDX), and fourier transform infrared (FTIR) spectra respectively. The luminescent behaviors of this phosphor were also investigated by photoluminescence (PL) and long afterglow (decay) characteristics.

EXPERIMENTAL

Material Preparation

The $Ca_2MgSi_2O_7:Eu^{3+}$ phosphor was prepared by the high temperature solid state reaction method. The raw materials are calcium carbonate [$CaCO_3$ (99.90%)], magnesium oxide [MgO (99.90%)], silicon di-oxide [SiO_2 (99.99%)] and europium oxide [Eu_2O_3 (99.99%)], all of analytical grade (A.R.), were employed in this experiment. Boric acid (H_3BO_3) was added as flux. Initially, the raw materials were weighed according to the nominal compositions of $Ca_2MgSi_2O_7:Eu^{3+}$ phosphor. Then the powders were mixed and milled thoroughly for 2 h using mortar and pestle. The grinded sample was placed in an alumina crucible and subsequently fired at 1200 °C for 3 h in air. At last the nominal compounds were obtained after the cooling down of programmable furnace.

Characterization Techniques

The phase structures of the prepared $Ca_2MgSi_2O_7:Eu^{3+}$ phosphor was characterized by powder X-ray diffraction analysis (XRD). XRD pattern has been obtained from Bruker D8 advanced X-ray powder diffractometer using CuKα radiation and the data were collected over the 2θ range 10–80°. Particle size of prepared phosphor was determined by TEM using TECHNAI G2. The samples required for TEM analysis were prepared by dispersing the sintered $Ca_2MgSi_2O_7:Eu^{3+}$ phosphor in methanol using an ultrasound bath technique. A drop of this dispersed suspension was put onto 200-mesh carbon coated copper grid and then dried into the air. An EDX spectra was used for the elemental analysis of the prepared phosphor (Sahu, Bisen, Brahme & Ganjir 2015c). FTIR spectra were recorded with the help of IR Prestige-21 by SHIMADZU for investigating the functional group ($4000-1400$ cm^{-1}) as well as the finger print region ($1400-400$ cm^{-1}) of sintered phosphor by mixing the sample with potassium bromide (KBr, AR grade). The PL measurements of excitation and emission spectra were recorded on a Shimadzu (RF 5301-PC) spectrofluorophotometer fitted with a sensitive photomultiplier tube. This spectrofluorophotometer provides corrected excitation and emission spectra in the 200–400 and 475–700 nm ranges, respectively. All measurements were carried out at the room temperature.

RESULTS AND DISCUSSION

XRD Analysis

In order to determine the phase structure, powder XRD analysis has been carried out. The typical XRD patterns of $Ca_2MgSi_2O_7$:Eu^{3+} phosphor with the standard XRD pattern is shown in Fig. 1. The position and intensity of diffraction peaks of the prepared $Ca_2MgSi_2O_7$: Eu^{3+} phosphor were matched and found to be consistent with the standard XRD pattern (COD card No. 96-900-6941) by MATCH 2 software. The figure of merit (FOM) while matching these was 0.9759 (97%) which illustrates that the phase of the prepared sample agrees with the standard pattern COD card No. 96-900-6941. In Fig. 1, it can be concluded that prepared samples are chemically and structurally $Ca_2MgSi_2O_7$ phosphor. The phase structure of the $Ca_2MgSi_2O_7$:Eu^{3+} phosphor is akermanite type structure which belongs to the tetragonal crystallography with space group $P\bar{4}2_1m$ (113 space number and D^3_{2d} space group), this structure is a member of the melilite group and forms a layered compound. The lattice parameters are calculated using Celref V3. The refined values of tetragonal europium doped di-calcium magnesium silicate were found as; a = b = 7.8470 Å, c = 5.0097 Å, $\alpha = 90°$, $\beta = 90°$, $\gamma = 90°$ and cell volume = 299.24 (Å)³, Z = 2 is nearly same [a = b = 7.8350 Å and c = 5.0100 Å, $\alpha = 90°$, $\beta = 90°$, $\gamma = 90°$ and cell volume = 299.36 (Å)³, Z = 2], with the standard lattice parameters which again signifies the proper preparation of the discussed $Ca_2MgSi_2O_7$:Eu^{3+} phosphor.

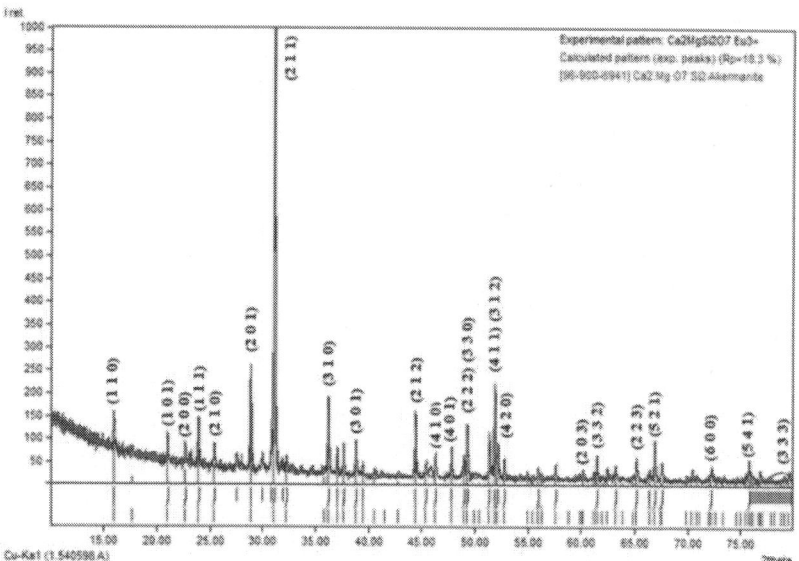

Figure 1. XRD patterns of $Ca_2MgSi_2O_7$:Eu^{3+} phosphor.

The average crystallite size was calculated from the XRD pattern using Debye Scherrer relation $D = k\lambda/\beta\cos\theta$, where D is the crystallite size for the (hkl) plane, λ is the wavelength of the incident X-ray radiation [CuKα (0.154 nm)], β is the full width at half maximum (FWHM) in radiations, and θ is the corresponding angle of Bragg diffraction. Sharper and isolated diffraction peaks such as $2\theta = 24.15$ (1 1 1), 29.04 (2 1 0), 31.26 (2 1 1), 36.44 (3 1 0), 38.93 (3 0 1), 44.55 (2 1 2) were chosen for calculation of the crystallite size. Based on the Debye-Scherrer's formula, the crystallite size is $\sim$ 73 nm, 70 nm, 68 nm, 69 nm, 67 nm, 66 nm was calculated, respectively and the average crystallite size is $\sim$68.83 nm.

Transmission Electron Microscopy (TEM)

Fig. 2 shows the transmission electron microscopy (TEM) image of $Ca_2MgSi_2O_7$:Eu^{3+} phosphor. From TEM image, it is seen that the, due the high temperature synthesis, the agglomeration of phosphor particles were observed. TEM images shows that the shape of the particle is tetragonal structure and particle size ranges in between 200 and 400 nm. So we conclude that, transmission electron microscopy results are in good agreement with the result of the XRD studies.

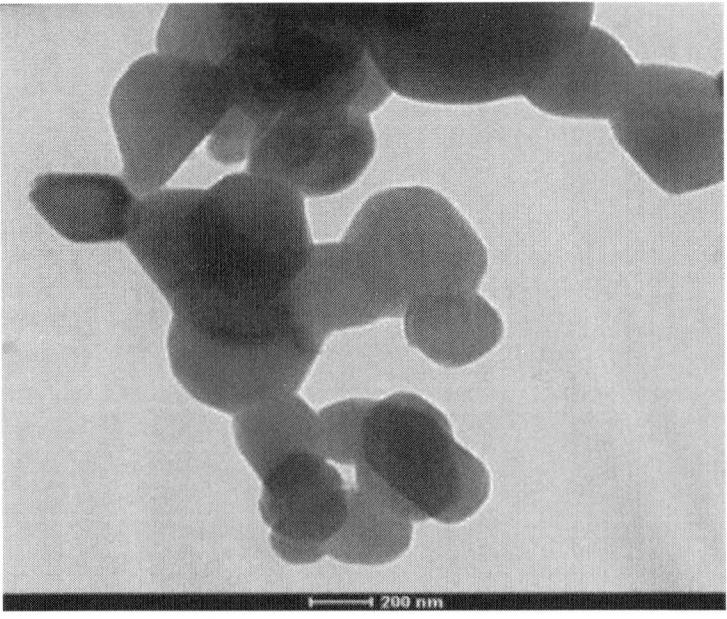

Figure 2. TEM image of $Ca_2MgSi_2O_7$:Eu^{3+} phosphor.

Field Emission Scanning Electron Microscopy (FESEM)

It is known that the luminescence characteristics of phosphor particles depend on the morphology of the particles, such as size, shape, size distribution, defects, and so on. The surface morphology of the $Ca_2MgSi_2O_7:Eu^{3+}$ phosphor is shown in Fig. 3(a, b) with different magnification. The surface morphology of the particles was not uniform and they aggregated tightly with each other. From the FESEM image, it can be observed that the prepared sample consists of particles with different size distribution. In addition, there are some big aggregates is also present due to high temperature heat treatment.

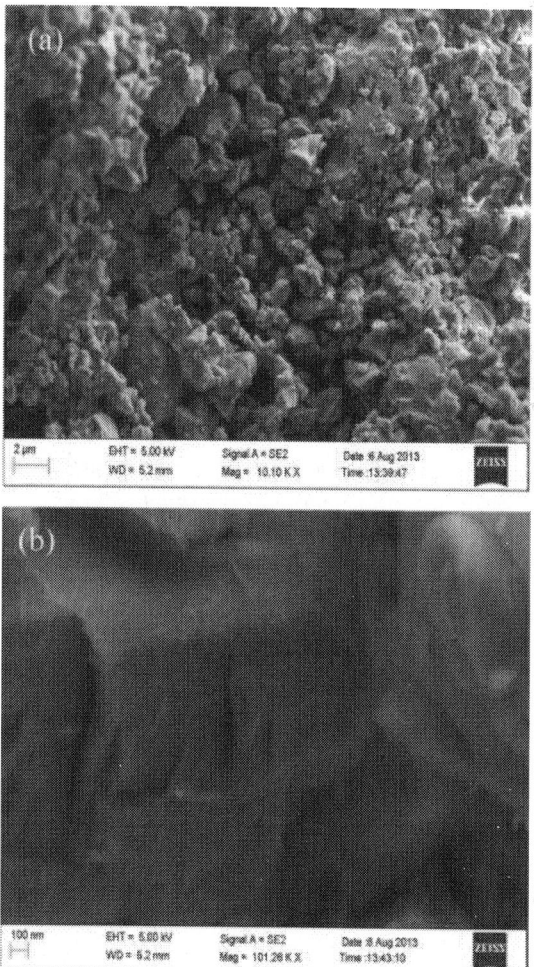

Figure 3. (a, b) FESEM image of $Ca_2MgSi_2O_7:Eu^{3+}$ phosphor.

Energy Dispersive X-Ray Spectroscopy (EDX)

The chemical composition of the powder sample has been measured using EDX spectra. EDX is a standard procedure for identifying and quantifying elemental composition of sample area as small as a few nanometers. The existence of europium (Eu) is clear in their corresponding EDX spectra. Their appeared no other emission apart from calcium (Ca), magnesium (Mg), silicon (Si) and oxygen (O) in $Ca_2MgSi_2O_7:Eu^{3+}$ EDX spectra of the phosphor. In EDX spectra, the presence of Ca, Mg, Si, O and Eu, intense peak are present which preliminary indicates the formation of $Ca_2MgSi_2O_7:Eu^{3+}$ phosphor in Fig. 4.

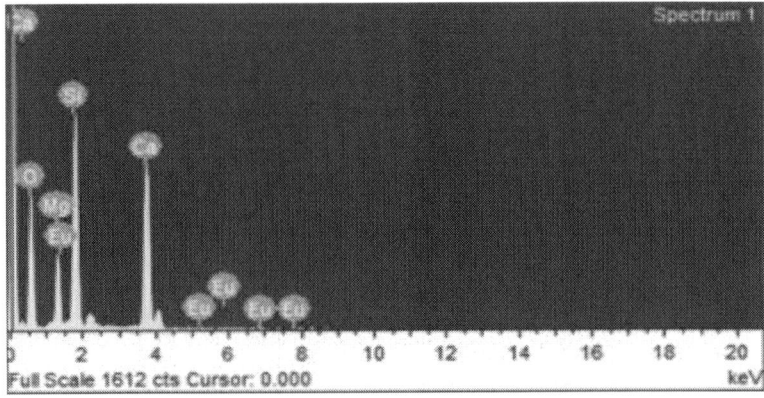

Figure 4. EDX spectra of $Ca_2MgSi_2O_7:Eu^{3+}$ phosphor.

Fourier Transform Infrared Spectra (FTIR)

The FTIR spectra has been widely used for the identification of organic and inorganic compounds. Fig. 5 shows the FTIR spectra of $Ca_2MgSi_2O_7:Eu^{3+}$ phosphor. In observed IR spectrum, the absorption bands of silicate groups were clearly evident. An intense band centred at 974.14 cm^{-1} is assigned due to Si–O–Si asymmetric stretch, bands at 646.29 cm^{-1} to Si–O symmetric stretch. Bands at 588.71 and 481.56 cm^{-1} are assigned to Si–O–Si vibrational mode. Furthermore, in keeping with the absorption bands, posited at 1010.92, 945.36 and 687.87 cm^{-1} can be ascribed to the presence of SiO_4 group. The band centred at 1783.38 cm^{-1} can be attributed to the presence of small amount of calcite (Gou, Chang, & Zhai, 2005).

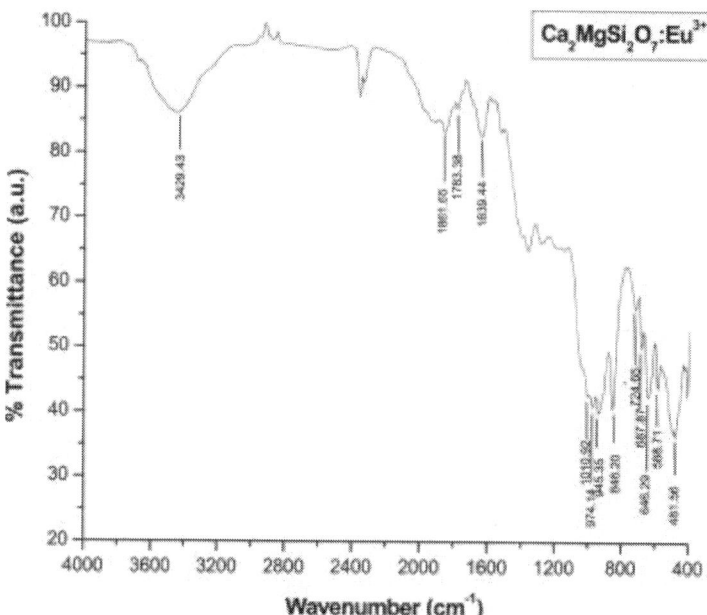

Figure 5. FTIR spectra of Ca₂MgSi₂O₇:Eu³⁺ phosphor.

The FTIR spectrum of Ca₂MgSi₂O₇:Eu³⁺ phosphor contain clearly exhibited bands in the region (3429.43 cm⁻¹) of hydroxyl group show the stretching vibration of O–H groups. The hydroxyl group in sintered phosphor is might be due to presence of moisture through environment. The asymmetric stretching of (CO_3^{2-}) carbonates can be observed in the range of 1900–1700 cm⁻¹ (Sahu, Bisen, & Brahme, 2014b). The weak shoulders, which corresponds to the out of plane bending of appears at ~1861.65 cm⁻¹. These bands are due to a slight carbonation of the samples preparation [CaCO₃ (raw material)]. The free CO_3^{2-} ions has a D3h symmetry (trigonal planar) and its spectrum is dominated by the band at 1900–1700 cm⁻¹. The vibration band of 1639.44 cm⁻¹ are assigned due to the Mg²⁺ and bending of the sharp peaks in the region of 846.20 and 724.65 cm⁻¹ are assigned due to Ca²⁺. When Eu³⁺ enters the lattice, it will replace the Ca²⁺ in the Ca₂MgSi₂O₇ host and occupy Ca²⁺ lattice sites due to distortion in the Ca₂MgSi₂O₇ host crystal lattice. Original position of Ca²⁺ was replaced by Eu³⁺ and the original of Ca²⁺located at somewhere. Therefore the vibration mode of Ca²⁺ at 846.20 and 724.65 cm⁻¹is clearly observed from $Ca_2MgSi_2O_7:Eu^{3+}$ phosphor (Chang and Mao, 2005 and Sahu et al., 2014c).

According to the crystal structure of $Ca_2MgSi_2O_7$, the coordination number of calcium can be 6 and 8. Therefore, Ca^{2+} can occupy two alternative lattice sites, the six coordinated Ca^{2+} site $[CaO_6$ (Ca (I) site)] and the eight coordinated Ca^{2+} site $[CaO_8$ (Ca (II) site)], and other two independent cations sites, namely Mg^{2+} $[MgO_4]$, and Si^{4+} $[SiO_4]$ also exist in the crystal lattice. Mg^{2+} and Si^{4+} cations occupy in the tetrahedral sites. Eu^{3+} ions can occupy with 3 oxidation state (3, 2 and 1) and five alternative lattice sites. The coordination number of europium can be 6, 7, 8, 9 and 10 (Eu (I), Eu (II), Eu (III), Eu (IV) and Eu (V), respectively) (Vicentini, Zinner, Zukerman-Schpector, & Zinner, 2000). It's hard for Eu^{3+} ions to incorporate the tetrahedral $[MgO_4]$ or $[SiO_4]$ symmetry but it can easily incorporate hexahedral $[CaO_6]$ or octahedral $[CaO_8]$. Another fact that supports that the radius of Eu^{3+} (1.07 Å) are very close to that of Ca^{2+} (about 1.12 Å) rather than Mg^{2+} (0.65 Å) and Si^{4+} (0.41 Å). Therefore, the Eu^{3+} ions are expected to occupy the Ca^{2+} sites in the $Ca_2MgSi_2O_7:Eu^{3+}$ phosphor (Chandrappa et al., 1999 and Salim et al., 2009).

Photoluminescence (PL)

The excitation spectrum of $Ca_2MgSi_2O_7:Eu^{3+}$ phosphor monitored at 593 nm emission is given in Fig. 6(a). The spectrum of $Ca_2MgSi_2O_7:Eu^{3+}$ phosphor exhibit a broad band in the UV region centered at about 265 nm, and several sharp lines between 300 to 400 nm. It can be seen from Fig. 6(a), the excitation spectrum is composed of two major parts: (1) the broad band between 220 and 300 nm, the broad absorption band is called charge transfer state (CTS) band due to the europium–oxygen interactions, which is caused by an electron transfer from an oxygen 2p orbital to an empty 4f shell of europium and the strongest excitation peak is at about 265 nm. (2) A series of sharp lines between 300 and 400 nm, ascribed to the f–f transition of Eu^{3+}. The strongest sharp peak is located at 395 nm corresponding to $^7F_0 \rightarrow {}^5L_6$ transition of Eu^{3+}. Other weak excitation peaks are located at 319, 363 and 383 nm are related to the intra-configurational 4f–4f transitions of Eu^{3+} ions in the host lattices, which can be assigned to $^7F_0 \rightarrow {}^5H_6$, $^7F_0 \rightarrow {}^5D_4$ and $^7F_0 \rightarrow {}^5L_7$ transitions,   respectively.   The prepared $Ca_2MgSi_2O_7$:Eu^{3+} phosphor can be excited by near UV (NUV) at about 395 nm effectively. So, it can match well with UV and NUV-LED, showing a great potential for practical applications (Wu et al., 2011 and Sahu et al., 2014d).

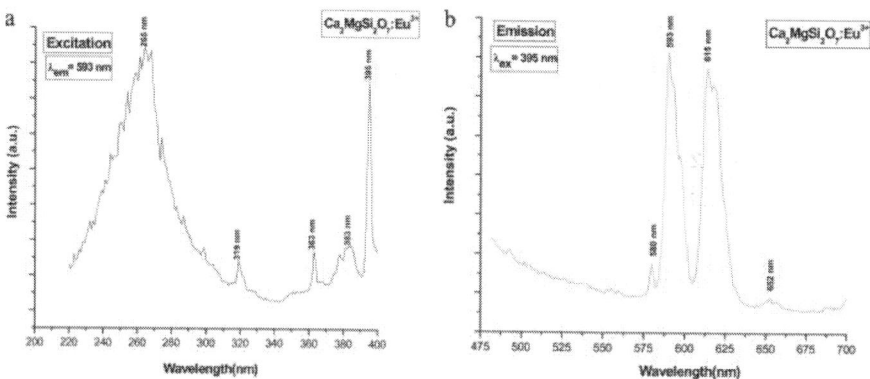

Figure 6. (a) Excitation spectra of $Ca_2MgSi_2O_7$:Eu^{3+} phosphor. (b) Emission spectra of $Ca_2MgSi_2O_7$:Eu^{3+} phosphor.

Fig. 6(b) shows the emission spectra of $Ca_2MgSi_2O_7$:Eu^{3+} phosphor in the range of 475–700 nm. Under the 395 nm excitation, the emission spectrum of our obtained samples was composed of a series of sharp emission lines, corresponding to transitions from the excited states $^5D_0$ to the ground state ^7F_j $(j = 0,1,2,3)$. The orange emission at about 593 nm belongs to the magnetic dipole $^5D_0 \rightarrow {}^7F_1$ transition of Eu^{3+}, and the transition hardly varies with the crystal field strength. The red emission at 615 nm ascribes to the electric dipole $^5D_0 \rightarrow {}^7F_2$ transition of Eu^{3+}, which is very sensitive to the local environment around the Eu^{3+}, and depends on the symmetry of the crystal field. It is found that the 593 and 615 nm emissions are the two strongest peaks, indicating that there are two Ca^{2+} sites in the $Ca_2MgSi_2O_7$ lattice. One site, Ca (I), is inversion symmetry and the other site, Ca (II), is non-inversion symmetry. When doped in $Ca_2MgSi_2O_7$ the Eu^{3+} ions occupied the two different sites of Ca (I) and Ca (II). Other two emission peaks located at 580 and 652 nm are relatively weak, corresponding to the $^5D_0 \rightarrow {}^7F_0$ and $^5D_0 \rightarrow {}^7F_3$ typical transitions of Eu^{3+} ions respectively (Kuang et al., 2014 and Sahu et al., 2014d).

For the phosphor $Ca_2MgSi_2O_7:Eu^{3+}$ prepared in our experiment, the strongest orange emission peak is located at 593 nm will be dominated. It can be presumed that Eu^{3+} ions mainly occupy with an inversion symmetric center in host lattice. Fig. 7 shows the schematic energy level diagram of Eu^{3+} ions in the $Ca_2MgSi_2O_7$ host depicting different emissions bands.

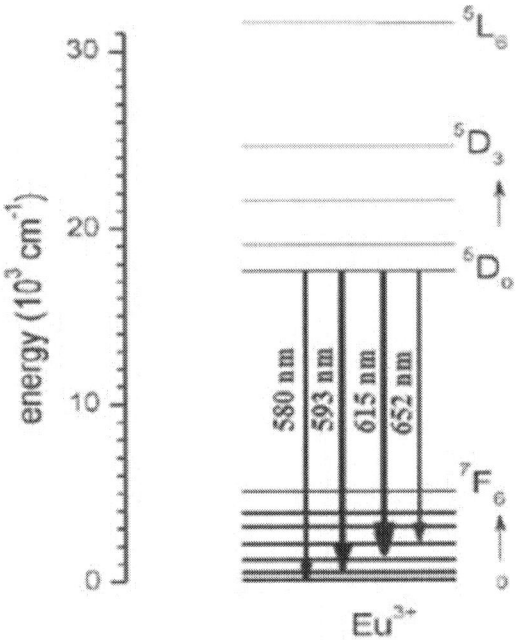

Figure 7. Schematic energy level diagram of $Ca_2MgSi_2O_7:Eu^{3+}$ phosphor.

Long Afterglow (Decay)

Fig. 8 shows the typical decay curves of $Ca_2MgSi_2O_7:Eu^{3+}$ phosphor. The initial afterglow intensity of the sample was high. The decay times of $Ca_2MgSi_2O_7:Eu^{3+}$ phosphor can be calculated by the curve fitting technique, and the decay curves fitted by the single exponential components have different decay times.

$$I = I_0 \exp(-t/\tau) \tag{1}$$

where, I_0 and I are the luminescence intensities at time 0 and t, respectively, and τ is the luminescence lifetime. Based on the decay curve and the above mentioned Eq. (1) the fitting curve result are shown in Table 1.

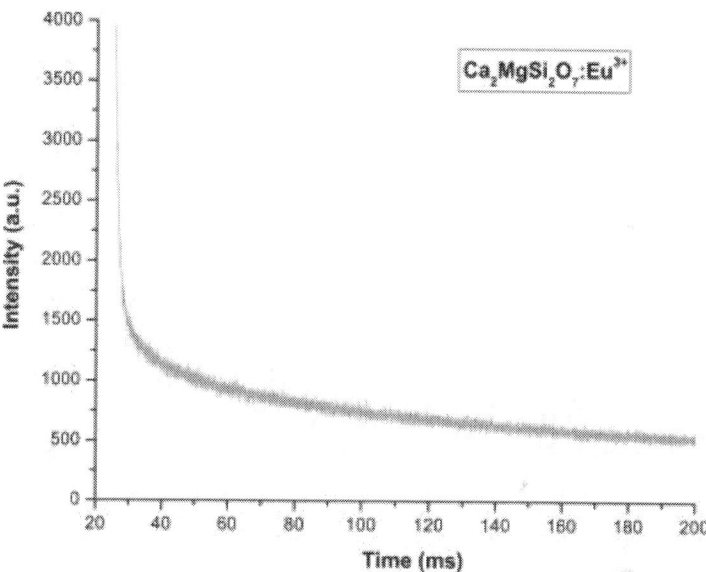

Figure 8. Decay Curve of $Ca_2MgSi_2O_7:Eu^{3+}$ phosphor.

Table 1. Fitting results of the decay curves.

Phosphor	τ (ms)
$Ca_2MgSi_2O_7:Eu^{3+}$	28.47

As it was reported before (Hong et al. 2011), when Eu^{3+} ions were doped into $Ca_2MgSi_2O_7$, they would substitute the Ca^{2+} ions. To keep electroneutrality of the compound, two Eu^{3+} ions would substitute three Ca^{2+} ions. The process can be expressed as

$$2Eu^{3+}+3Ca^{2+}\rightarrow 2[Eu_{Ca}]^{*}+[V_{Ca}]''$$

Each substitution of two Eu^{3+} ions would create two positive defects of $[Eu_{Ca}]^{*}$ capturing electrons and one negative vacancy of $[V_{Ca}]''$. These defects act as trapping centers for charge carriers. Then the vacancy $[V_{Ca}]''$ would act as a donor of electrons while the two $[Eu_{Ca}]^*$ defects become acceptors of electrons. By thermal stimulation, electrons of the $[V_{Ca}]''$ vacancies would then transfer to the Eu^{3+} sites. The results indicate that the depth of the trap is too shallow leading to a quick escape of charge carriers from the traps resulting in a fast recombination rate in milliseconds (ms).

CIE Chromaticity Coordinate

In general, color of any phosphor material is represented by means of color coordinates. The luminescence color of the samples excited under 395 nm has been characterized by the CIE (Commission International de I'Eclairage) 1931 chromaticity diagram. The emission spectrum of the Eu^{3+} doped $Ca_2MgSi_2O_7$ phosphor was converted to the CIE 1931 chromaticity using the photoluminescent data and the interactive CIE software (CIE coordinate calculator) diagram as shown in Fig. 9 (Sahu, Bisen, & Brahme, 2014a).

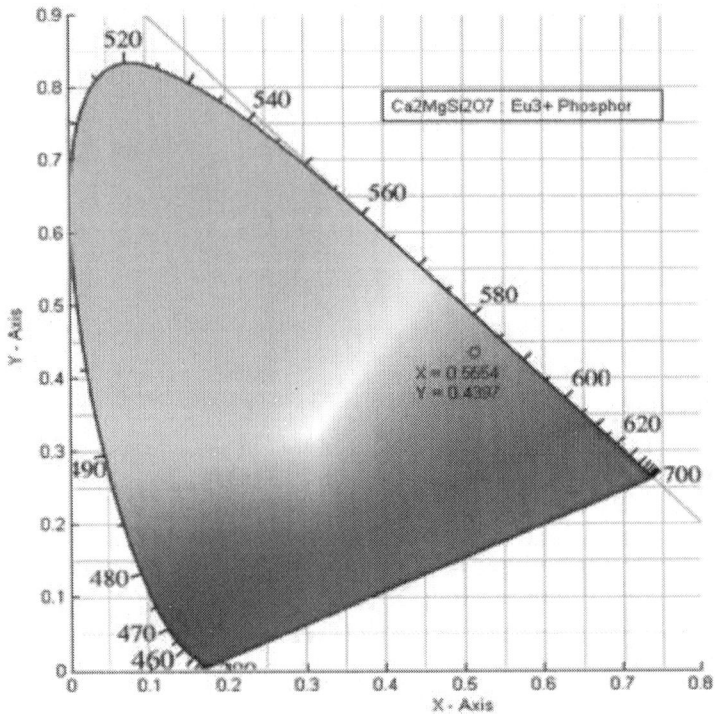

Figure 9. CIE chromaticity diagram of $Ca_2MgSi_2O_7$:Eu^{3+} phosphor.

Every natural color can be identified by (X, Y) coordinates that are disposed inside the 'chromatic shoe' representing the saturated colors. Luminescence colors of Eu^{3+} doped $Ca_2MgSi_2O_7$ phosphor are placed in the orange-red (X = 0.5554, Y = 0.4397) corners. The chromatic coordinates of the luminescence of this phosphor are measure and reached to orange-red luminescence (CIE 1931).

CONCLUSION

A new orange–red emitting phosphor of $Ca_2MgSi_2O_7$:Eu^{3+} was synthesized by high temperature solid state reaction method at 1200 °C and its photoluminescence properties were investigated. The EDX spectra confirm the present elements in $Ca_2MgSi_2O_7$:Eu^{3+} phosphor. The excitation spectra indicate the phosphor can be effectively excited by near ultraviolet (NUV) light, making it attractive as conversion phosphor for LED applications. The $Ca_2MgSi_2O_7$:Eu^{3+} phosphor exhibits bright orange–red emission excited by 395 nm. Photoluminescence measurements showed that the phosphor exhibited emission peak with good intensity at 593 nm, corresponding to $^5D_0 \rightarrow {}^7F_1$ orange emission and weak $^5D_0 \rightarrow {}^7F_2$ red emission. The band at 395 nm can be assigned to $^7F_0 \rightarrow {}^5L_6$ transition of Eu^{3+} ions due to the typical f–f transitions within Eu^{3+} of $4f^6$ configuration. CIE chromaticity diagram confirms $Ca_2MgSi_2O_7$:Eu^{3+} phosphor exhibits efficient orange–red emission and excellent color stability, indicating that it has favorable properties for application as near ultraviolet LED conversion phosphor.

ACKNOWLEDGMENT

"We are very grateful to UGC-DAE Consortium for Scientific Research, Indore (M.P.) for the XRD Characterization and we are very thankful Dr. Mukul Gupta for his co-operation". We are very thankful to Dr. K.V.R. Murthy, Department of Applied physics, M.S. University Baroda, Vadodara (Gujarat) India for the photoluminescence study.

REFERENCES

1. Chandrappa, G. T., Ghosh, S., & Patil, K. C. (1999). Synthesis and properties of Willemite, Zn2SiO4, and M2þ:Zn2SiO4 (M ¼ Co and Ni). Journal of Materials Synthesis and Processing, 7, 273e279.
2. Chang, C., & Mao, D. (2005). Luminescent properties of Sr2MgSi2O7 and Ca2MgSi2O7 long lasting phosphors activated by Eu2 þ, Dy3 þ. Journal of Alloy and Compound, 390, 133e137.
3. CIE (1931). International Commission on Illumination. Publication CIE no. 15 (E-1.3.1).
4. Dong, X., Zhang, J., Zhang, X., Hao, Z., & Luo, Y. (2014). New orangeered phosphor Sr9Sc (PO4)7:Eu3þ for NUV-LEDs application. Journal of Alloys and Compounds, 587, 493e496.
5. Gorller-Walrand, C., Fluyt, L., Ceulemans, A., & Carnall, W. T. (1995). Magnetic dipole transitions as standards for JuddeOfelt parametrization in lanthanide spectra. Journal of Chemical Physics, 1991, 3099e3106.
6. Gou, Z., Chang, J., & Zhai, W. (2005). Preparation and characterization of novel bioactive dicalcium silicate ceramics. Journal of the European Ceramic Society, 25, 1507e1514.
7. Hong, Y., Guimei, G., Li, K., Guanghuan, L., Shucai, G., & Guangyan, H. (2011). Synthesis and luminescence properties of a novel red-emitting phosphor SrCaSiO4:Eu3þ for ultraviolet white light-emitting diodes. Journal of Rare Earths, 29(5), 431e435.
8. Jiao, H. Y., & Wang, Y. (2012). A potential red-emitting phosphor CaSrAl2SiO7:Eu3þ for near ultraviolet light emitting diodes. Physica B, 407, 2729e2733.
9. Kuang, S. P., Liang, K., Liu, J., Mei, Y. M., Jiang, M., Wu, Z. C., et al. (2014). Preparation and photoluminescence properties of a new orangeered Ba3P4O13:Eu3þ phosphor. Optik e International Journal for Light and Electron Optics. http://dx.doi.org/10.1016/ j.ijleo.2013.12.037.
10. Liu, J., Liang, K., Wu, Z. C., Mei, Y. M., Kuang, S. P., & Li, D. X. (2014). The reduction of Eu3þ to Eu2þ in a new orangeered emission Sr3P4O13:Eu phosphor prepared in air and its photoluminescence properties. Ceramics International, 40, 8827e8831.
11. Sahu, I. P., Bisen, D. P., & Brahme, N. (2014a). Dysprosium doped di-strontium magnesium di-silicate White light emitting phosphor by solid state reaction method. Displays, 35, 279e286.
12. Sahu, I. P., Bisen, D. P., & Brahme, N. (2014b). Structural characterization and optical properties of Ca2MgSi2O7:Eu2þ,Dy3þ phosphor by solid-state reaction method. Luminescence:The Journal of Biological and Chemical Luminescence. http://dx.doi.org/10.1002/bio.2771 (Wiley publication).

13. Sahu, I. P., Bisen, D. P., & Brahme, N. (2014c). Luminescence properties of Eu2þ and Dy3þ doped Sr2MgSi2O7 and Ca2MgSi2O7 phosphors by solid state reaction method. Research on Chemical Intermediates. http://dx.doi.org/10.1007/s11164-014-1767-6

14. (Springer publication).

15. Sahu, I. P., Bisen, D. P., & Brahme, N. (2014d). Photoluminescence properties of europium doped di-strontium magnesium disilicate phosphor by solid state reaction method. Journal of Radiation Research and Applied Sciences, 8(1), 104e109. http:// dx.doi.org/10.1016/j.jrras.2014.12.006.

16. Sahu, I. P., Bisen, D. P., & Brahme, N. (2015a). Luminescence properties of Green emitting Ca2MgSi2O7:Eu2þ phosphor by solid state reaction method. Lumiescence: The Journal of Biological and Chemical Luminescence. http://dx.doi.org/10.1002/ bio.2869 (Wiley publication).

17. Sahu, I. P., Bisen, D. P., Brahme, N., Wanjari, L., & Tamrakar, R. K. (2015b). Structural characterization and luminescence properties of Bluish-Green emitting SrCaMgSi2O7:Eu2þ, Dy3þ phosphor by solid state reaction method. Research on Chemical Intermediates. http://dx.doi.org/10.1007/s11164-015-1929-1 (Springer publication).

18. Sahu, I. P., Bisen, D. P., Brahme, N., & Ganjir, M. (2015c). Enhancement of the photoluminescence and long afterglow properties of Sr2MgSi2O7:Eu2þ phosphor by Dy3þ co-doping. Lumiescence: The Journal of Biological and Chemical Luminescence. http://dx.doi.org/10.1002/bio.2900 (Wiley publication).

19. Salim, M. A., Hussain, R., Abdullah, M. S., Abdullah, S., Alias, N. S., & Ahmad Fuzi, S. A. (2009). The local structure of phosphor material, Sr2MgSi2O7 and Sr2MgSi2O7:Eu2þ by infrared spectroscopy. Solid State Science and Technology, 17, 59e64.

20. Shrivastava, R., & Kaur, J. (2014). Studies on long lasting optical properties of europium and dysprosium doped di-strontium magnesium silicate phosphors. Indian Journal of Physics. http:// dx.doi.org/10.1007/s12648-014-0535-1.

21. Talwar, G. J., Joshi, C. P., Moharil, S. V., Dhopte, S. M., Muthal, P. L., & Kondawat, V. K. (2009). Combustion synthesis of Sr3MgSi2O8:Eu2þand Sr2MgSi2O7:Eu2þphosphors. Journal of Luminescence, 129 (11), 1239.

22. Vicentini, G., Zinner, L. B., Zukerman-Schpector, J., & Zinner, K. (2000). Luminescence and structure of europium compounds. Coordination Chemistry Reviews, 196, 353e382.

23. Wang, Z., Lou, S., & Li, P. (2014). Enhanced orangeered emission of Sr3La (PO4)3:Ce3þ ,Mn2þ via energy transfer. Journal of Luminescence, 156, 87e90.

24. Wu, H., Hu, Y., Wang, Y., Kang, F., & Mou, Z. (2011). Investigation on Eu3þ doped Sr2MgSi2O7 red emitting phosphors for white light emitting diodes. Optics & Laser Technology, 43, 1104e1110.
25. Xu, M., Wang, L., Liu, L., Jia, D., & Sheng, R. (2014). Influence of Gd3þ doping on the luminescent of Sr2P2O7:Eu3þ orangeered phosphors. Journal of Luminescence, 146, 475e479.
26. Yang, H. K., Moona, B. K., Choi, B. C., Jeong, J. H., & Kim, K. H. (2012). Crystal growth and photoluminescence characteristics of Ca2MgSi2O7:Eu3þ thin films grown by pulsed laser deposition. Materials Research Bulletin, 47, 2871e2874.

CITATION

Ishwar Prasad Sahu, D.P. Bisen, Nameeta Brahme, Europium doped di-calcium magnesium di-silicate orange–red emitting phosphor by solid state reaction method, Journal of Radiation Research and Applied Sciences, Volume 8, Issue 3, July 2015, Pages 381-388, ISSN 1687-8507, http://dx.doi.org/10.1016/j.jrras.2015.02.007.

CHAPTER 11

Sensitivity Improvement During Heteronuclear Spin Decoupling in Solid-State Nuclear Magnetic Resonance Experiments at High Spinning Frequencies and Moderate Radio-Frequency Amplitudes

Rudra N. Purusottam[1, 2, 3], Geoffrey Bodenhausen[1, 2, 3] and Piotr Tekely[1, 2, 3]

[1] École Normale Supérieure – PSL Research University, Département de Chimie, 24, rue Lhomond, F-75005 Paris, France

[2] Sorbonne Universités, UPMC University Paris 06, LBM, 4 place Jussieu, F-75005 Paris, France

[3] CNRS, UMR 7203 LBM, F-75005 Paris, France

ABSTRACT

Searching for optimal conditions during one- and multi-dimensional solid-state NMR experiments in high static fields may require spinning the sample at frequencies above 40 kHz. This implies challenging requirements for heteronuclear spin decoupling. We have compared the performance of the latest heteronuclear decoupling schemes at high magic-angle spinning frequencies. The results demonstrate that at commonly used rf amplitudes between 80 and 120 kHz, PISSARRO decoupling provides substantial sensitivity improvement. The performance of low-amplitude decoupling at different spinning speeds is also compared and its dependence on the inherent inhomogeneity of the rf field is probed by numerical simulations.

INTRODUCTION

Efficient heteronuclear decoupling is vital for obtaining high-resolution solid-state NMR spectra of low-gamma nuclei such as carbon-13. In polycrystalline and amorphous organic solids studied at magic-angle-spinning (MAS) frequencies above 5 kHz, flip-flop spin exchange between protons slows down and the efficiency of continuous-wave (CW) decoupling is not sufficient [1]. This disadvantage can be overcome by substituting CW irradiation by phase-alternated irradiation [1]. This was followed by the popular two-pulse phase-modulated (TPPM) technique [2] and its numerous variants [3], [4], [5], [6], [7] and [8] that have been successfully used for common spinning frequency ranges between 10 and 30 kHz. At high static fields, higher spinning frequencies may be preferred to attenuate residual spinning sidebands and unwanted rotational resonance effects that occur when an integer multiple of the spinning frequency ν_{rot} is roughly matched with the difference $\Delta\nu_{iso}$ between two isotropic chemical shifts ($n\nu_{rot} = \Delta\nu_{iso}$) [9]. This may lead to harmful line broadening or to undesirable magnetization exchange between specific sites. High spinning frequencies may also be useful at high static fields to create optimal conditions for broadband magnetization exchange in two-dimensional homonuclear correlation experiments [10]. Other indirect benefits of high spinning frequencies are related to the use of small-diameter rotors that allow one to run experiments with less than 1 mg powder sample. However, spinning frequencies around and above 30 kHz may also lead to a dramatic breakdown of the decoupling efficiency over a large range of rf amplitudes due to the phenomenon of rotary resonance recoupling (R^3) ($\nu_1 = n\nu_{rot}$) [11]. To overcome this complication, a phase-inverted supercycled sequence for attenuation of rotary resonance (PISSARRO) was developed and shown to be effective in quenching rotary resonance recoupling in the vicinity of $n = 2$ [12]. The method turned out also to achieve a very good decoupling efficiency at high rf amplitudes, far from any R^3 condition, as well as in the low-amplitude decoupling regime when $\nu_{rot} = 60$ kHz [13]. This is partially related to its capacity to make use of the modulation sidebands, which arise from the interference between the decoupling irradiation and the modulation of dipolar couplings by MAS [12]. A thorough analysis of the mechanism of quenching of rotary resonance recoupling effects by the

PISSARRO scheme has revealed the crucial role of its mirror symmetry segments combined with phase-shifted irradiation [14]. The immunity of PISSARRO decoupling against the offsets of remote protons, their chemical shift anisotropies and second-order cross-terms between dipolar coupling and chemical-shift anisotropy has also been demonstrated [14] and [15]. Since the introduction of PISSARRO decoupling, a new class of pulse sequences, so-called refocused continuous-wave (rCW), has been introduced [16]. Although rCW pulse sequences were so far only used for spinning frequencies up to 20 kHz, it has been suggested that they would also be efficient at high spinning speeds [17]. Independently, an amplitude-modulated XiX irradiation was proposed for the low-amplitude decoupling regime and tested at spinning frequencies of 60 and 90 kHz [18].

In this work, we show that at high spinning speeds, PISSARRO irradiation has a unique capacity to secure efficient decoupling at commonly used rf amplitudes between 80 and 120 kHz and provides significant sensitivity improvements. We also compare the efficiency of low-amplitude decoupling for lower spinning frequencies and point out the dependence of decoupling on the rf field inhomogeneity of the solenoid coil. For the sake of comparison, we also refer to the decoupling performance of SWf-TPPM [6], which according to recent reports [19], offers improved efficiency over a larger range of rf amplitudes and spinning frequencies compared to its precursors.

EXPERIMENTAL

Polycrystalline powders of uniformly ^{13}C, ^{15}N-labeled L-histidine hydrochloride monohydrate were used without further purification. All experiments were performed on a 400 MHz Bruker Avance II spectrometer and on a 850 MHz Bruker Avance III spectrometer, both equipped with double resonance CP/MAS probes using rotors with 1.3 mm diameter. For PISSARRO decoupling, only the pulse duration τp needs to be optimized in the vicinity of the recommended values, i.e. $\tau p = 0.2\tau_{rot}$ for decoupling near the $n = 2$ rotary resonance condition, and $\tau p = 0.9$ or $1.1\tau_{rot}$ for high rf amplitudes $v_1^H \gg 2v_{rot}$ [12]. For low-amplitude decoupling, the pulse lengths were optimized so that the nutation angles $\beta = 2\pi v_1 \tau$ were in the vicinity of 6π [13]. For routine purposes, the pulse duration τp need not be optimized and may be safely fixed to the above-mentioned durations, depending on the spinning frequency and rf amplitude. This gives a major advantage of PISSARRO over current decoupling schemes. For SWf-TPPM decoupling [6], optimum nutation angles were found in the range $120° < \beta < 170°$. The phase angle ϕ was optimized between $5°$ and $20°$ in steps of $2°$. The number of pulse pairs was fixed to 11 and a linear sweep profile was applied. For rCW decoupling, we chose the rCWc version [16] for best performance under our experimental conditions and lower demands on rf field strengths. For the refocusing π pulses, an rf field amplitude of 220 kHz was used. The optimum continuous wave pulse length, i.e. the delay τ between the refocusing pulses, was around 50 μs ($\tau = n\tau_{rot}, n \cong 3$). For low-amplitude experiments recorded with the basic AM-XiX scheme [18], the XiX component was optimized empirically for various MAS frequencies ($v_1^H = 11.9, 10.1, 8.6$ kHz for $v_{rot} = 60, 50$ and 40 kHz MAS, respectively). The rf amplitude of the continuous wave component, as defined in [18], ranged between 2 and 3 kHz. The pulse length was optimized around $\tau p \cong 5/2\tau r$. Numerical simulations were carried out with SPINEVOLUTION [20], considering the 5-spin system $C^\alpha H^\alpha H^{\beta 1} H^{\beta 2} H^N$ of L-histidine hydrochloride monohydrate with internuclear distances derived from the crystallographic structure. In each case the pulse durations were optimized numerically in the vicinity of the recommended values.

RESULTS AND DISCUSSION

The performance of PISSARRO, SWf-TPPM and rCWc at $v_{rot} = 40$ and 60 kHz in a medium static magnetic field ($B_0 = 9.4$ T) is compared in Figure 1 and Figure 2 for all carbon-13 resonances of L-histidine.

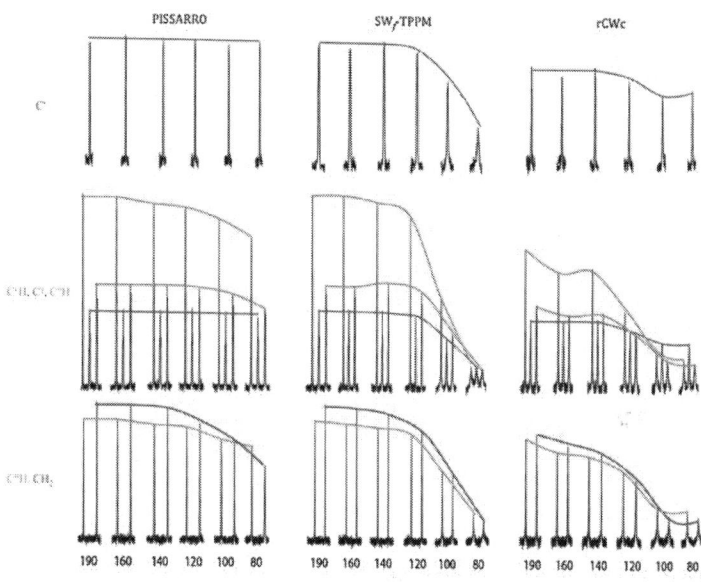

Figure 1. Comparison of the efficiency of heteronuclear decoupling for different carbons in L-histidine with PISSARRO (left), SW$_f$-TPPM (middle) and rCWc (right) at $B_0 = 9.4$ T (400 MHz for protons) and $v_{rot} = 40$ kHz as a function of the decoupling *rf* amplitude $190 > v_1 > 80$ kHz with the ^{1}H carrier frequency placed on-resonance for C$^\alpha$H. All spectra were recorded with 3.0 ms CP contact time, 8 scans and a 5 s recovery delay between experiments.

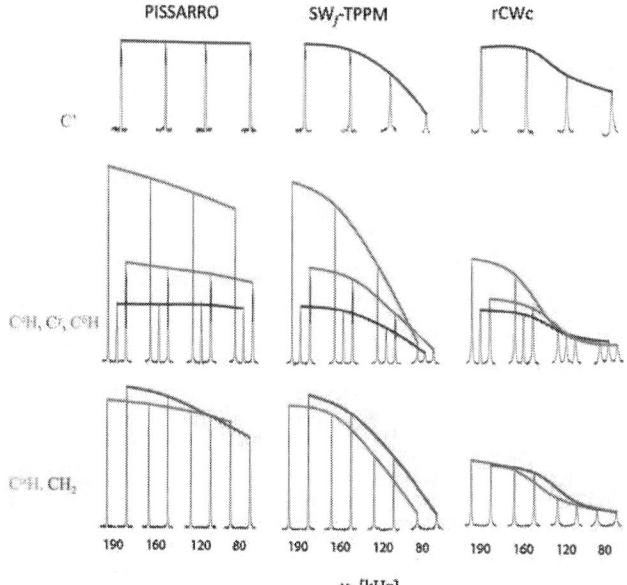

Figure 2. Same as in Figure 1 except that $v_{rot} = 60$ kHz.

A wide range of rf decoupling amplitudes $190 > v_1^H > 80$ kHz was tested. For 40 kHz MAS and high rf amplitudes between 190 and 120 kHz, PISSARRO and SWf-TPPM offer similar performance, while lower peak heights are observed for rCWc. In contrast to PISSARRO, the other two methods show a significant drop of performance for $v_1^H < 120$ kHz. This drop is greatly amplified at $v_{rot} = 60$ kHz due to the destructive interference near the $n = 1$ and 2 rotary resonance conditions.

Although somewhat better decoupling for $v_1^H < 120$ kHz can be achieved with high-phase TPPM [21] than with SWf-TPPM at $v_{rot} = 60$ kHz, the peak heights are only about 60–80% of those obtained with PISSARRO [15].

In ^{13}C spectra of organic solids recorded at high spinning frequencies and very high rf decoupling amplitudes, the observed line-widths are often not only determined by the performance of decoupling, but may be governed by the inhomogeneous distribution of isotropic chemical shifts and, to a lesser extent, by magnetic susceptibility, dynamic effects and minor instrumental misadjustments. By combining heteronuclear decoupling with spin echoes, one can determine the residual homogeneous line-width (also known as 'refocused line-width' characterized by the time constant T_2') that is relevant for many multipulse and multidimensional experiments [22] and [23]. To check to what extent T_2' is affected by the rf amplitude, we measured the echo decays of ^{13}C$^\varepsilon$H, which is not involved in homonuclear J_{CC} couplings and has a total line-width $\delta H^*_{1/2} = 52.6$ Hz at $v_1^H = 160$ kHz. The longest ^{13}C T_2' is observed with SWf-TPPM at $v_1^H = 160$ kHz. However, this does not translate into the highest peak of the ^{13}C$^\varepsilon$H resonance which is merely 86% of the peak height obtained with PISSARRO at the same decoupling amplitude. This corroborates earlier observations that the peak-heights and the line-widths are not simply inversely proportional to each other [14] and [22]. As shown in Figure 3, a significant drop of T_2' is observed with SWf-TPPM when decreasing the rf amplitude. For $v_1^H = 160$, 120 and 80 kHz, we measured $T_2' = 66.7$, 26.3 and 8.7 ms, respectively. This translates into homogeneous line-widths at half-height $\delta H'_{1/2} = 4.8$, 12.1 and 36.6 Hz. The corresponding total line-widths with SWf-TPPM were 53.5, 55.7 and

318.6 Hz. In contrast, the lifetimes measured with PISSARRO are largely independent of v_1^H ($T_{2'}$ = 43.5, 41.7 and 37.0 ms for v_1^H = 160, 120 and 80 kHz, respectively). These lifetimes provide satisfactory sensitivity without resorting to high rf amplitude. The corresponding homogeneous line-widths are 7.3, 7.6 and 8.6 Hz while the total line-widths are 52.6, 54.1 and 56.1 Hz.

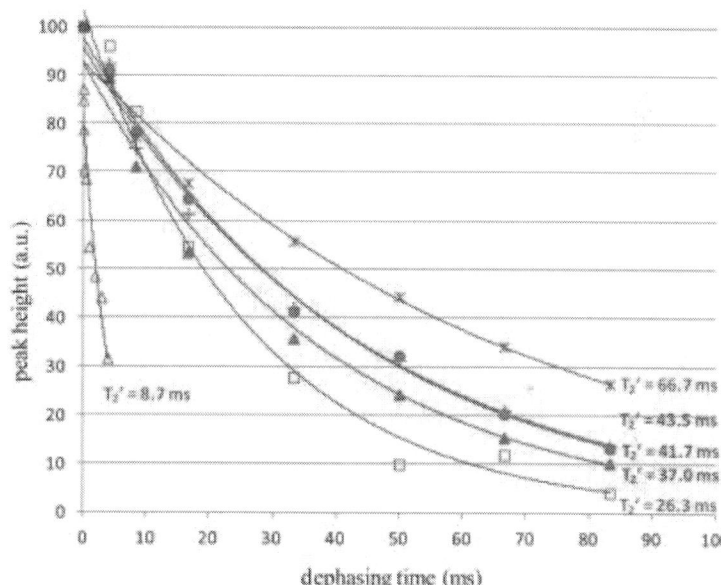

Figure 3. Echo decay curves for the $C^\varepsilon H$ carbon of L-histidine obtained using a single rotor-synchronized refocusing pulse at B_0 = 9.4 T and v_{rot} = 60 kHz. The SW$_f$-TPPM pulse sequence was used with rf decoupling amplitude of 160 (fx2), 120 (fx3) and 80 kHz (fx4). The T_2' relaxation times are respectively 66.7, 26.3 and 8.7 ms. The decays for PISSARRO at the same rf amplitudes are represented by red symbols (fx5, fx6 and fx7) and the corresponding T_2' values are 43.5, 41.7 and 37.0 ms. (For interpretation of the references to color in this figure legend, the reader is referred to the web version of this article.)

Figure 4 shows the decoupling efficiency of PISSARRO, SWf-TPPM and rCWc in high static magnetic field (B_0 = 19.9 T, or 850 MHz for protons) and v_{rot} = 60 kHz. Similar changes in decoupling performance as a function of v_1^H as observed at low field (Figure 2) occur, except that the efficiency of SWf-TPPM is somewhat improved near the n = 2 rotary resonance condition. The data corroborate a substantial sensitivity improvement with PISSARRO decoupling for commonly used rf amplitudes 80 < v_1^H < 120 kHz over a wide range of static fields.

This will help in high MAS experiments specifically designed to enhance the sensitivity, e.g., by promoting uniform Overhauser enhancements in ^{13}C NMR spectra of microcrystalline proteins [24].

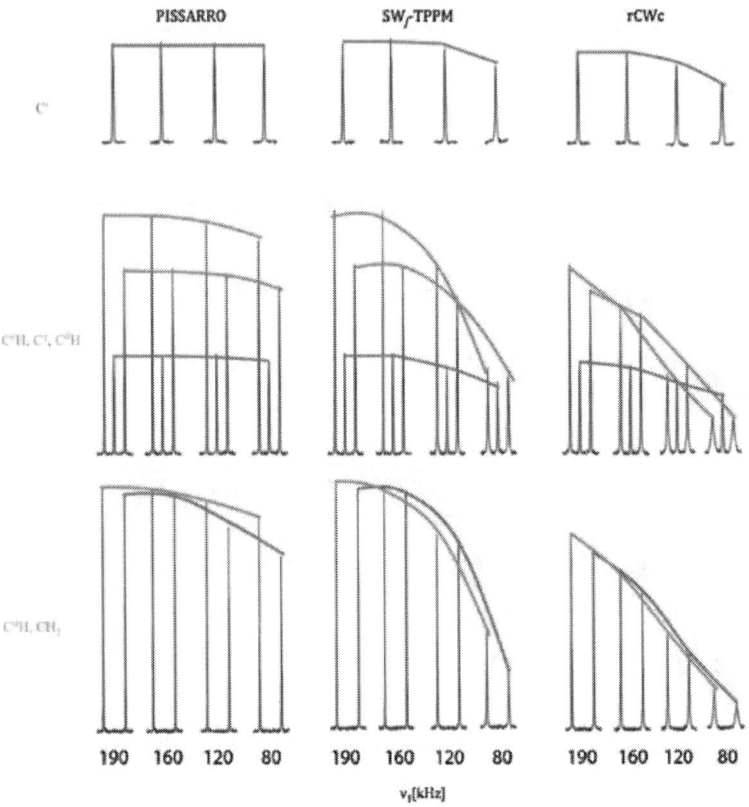

Figure 4. Comparison of the efficiency of heteronuclear decoupling for different carbons in L-histidine with PISSARRO (left), SW$_f$-TPPM (middle) and rCWc (right) at $B_0 = 19.9$ T (850 MHz for protons) and $v_{rot} = 60$ kHz as a function of the decoupling *rf* amplitude. All spectra were recorded after a pre-saturation of ^{1}H during 15.0 ms, 5 s relaxation delay, 1.3 ms CP contact time, 8 scans and 5 s delays between experiments.

Heteronuclear spin decoupling can also be efficient in a low *rf* amplitude regime with $v_1^H < 30$ kHz provided that $v_{rot} > 40$ kHz [7], [13], [18], [25], [26] and [27]. This allows a substantial reduction of the *rf* power dissipation, which may be vital for heat-sensitive samples. It has been demonstrated earlier that at $v_{rot} = 60$ kHz in a very high magnetic field $B_0 = 21$ T (900 MHz for protons), PISSARRO decoupling with $v_1^H = 15$ kHz has nearly the same efficiency as with $v_1^H = 150$ kHz for both CH_3 and CH resonances in alanine, while the peak height of CH_2 in glycine was 20% lower [13]. The comparison of peak heights shown in Figure 5 for L-histidine recorded at $B_0 = 19.9$ T and $v_{rot} = 60$ kHz in the high- and low-amplitude decoupling regimes reveals similar losses for various decoupling schemes.

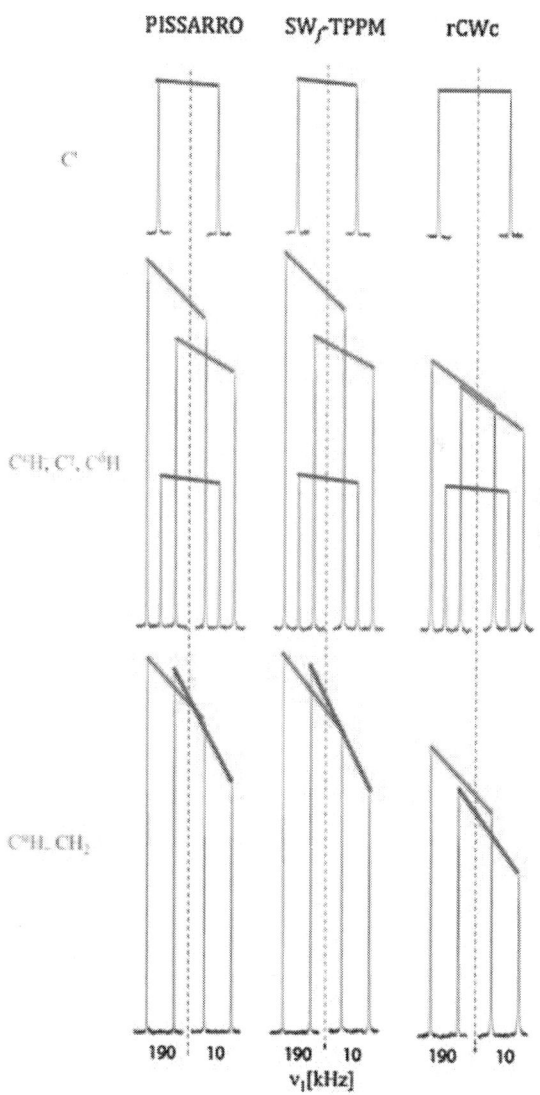

Figure 5. Comparison of the efficiency of heteronuclear decoupling for different carbons in L-histidine with PISSARRO (left), SW$_f$-TPPM (middle) and rCWc (right) at $B_0 = 19.9$ T and $v_{rot} = 60$ kHz in the high- and low-amplitude decoupling regime. Other parameters as in Figure 4.

For the sake of completeness, Figure 6 shows the peak heights of L-histidine recorded with PISSARRO, SWf-TPPM and AM-XiX in the low amplitude regime ($10 < v_1^H < 11$ kHz) at $B_0 = 9.4$ T and at three spinning frequencies. The performance is nearly the same for all three pulse sequences, though slightly better for PISSARRO at $v_{rot} = 40$ kHz. More importantly, the observed loss of peak heights at lower spinning frequencies compels one to resort to rf amplitudes in the range of $80 < v_1^H < 100$ kHz where PISSARRO offers the best efficiency among current decoupling schemes.

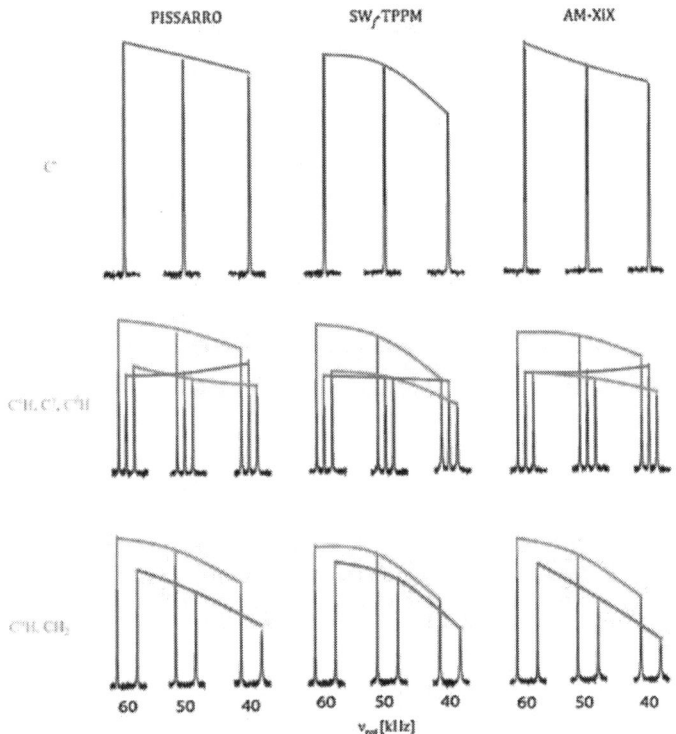

Figure 6. Comparison of the efficiency of low-amplitude heteronuclear decoupling for different carbons in L-histidine with PISSARRO (left), SWf-TPPM (middle) and AM-XiX (right) at $B_0 = 9.4$ T and three spinning frequencies v_{rot}. The optimized decoupling rf amplitudes ranged between 10 and 11 kHz to get the best efficiency in each case. All spectra were recorded with a single pulse experiment preceded by pre-saturation of ^{13}C during 15.0 ms and 5 s relaxation delay, 8 scans and 3 s delay between experiments. The observed increase of the intensity of a quaternary aromatic C^γ carbon at the lowest spinning frequency results from its faster longitudinal relaxation.

All decoupling sequences use pulse durations that must be optimized for a given *rf* power. However, there is an unavoidable distribution of *rf* field amplitudes, mostly along the axis of the solenoid coil. To assess the consequences of this distribution on the performance of heteronuclear decoupling, the $^{13}C^{\alpha}H$ resonance line was calculated numerically, again considering a spin cluster $C^{\alpha}H^{\alpha}H^{\beta 1}H^{\beta 2}H^{N1}$ in L-histidine with high- and low-amplitude PISSARRO decoupling at 9.4 and 19.9 T and $\nu_{rot} = 60$ kHz (Figure 7).

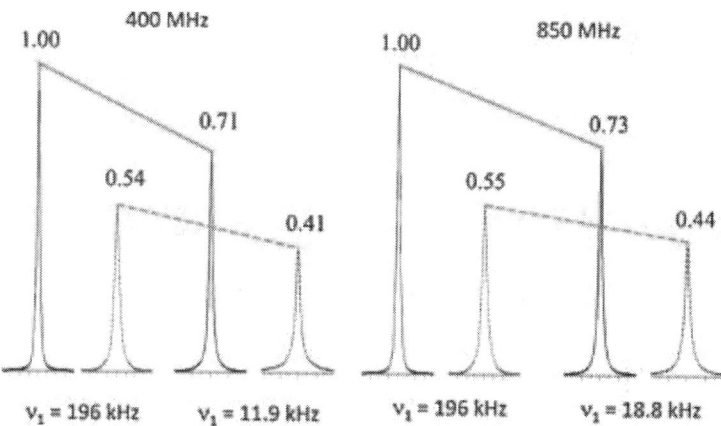

Figure 7. Numerically calculated $^{13}C^{\alpha}H$ lines, considering a spin cluster $C^{\alpha}H^{\alpha}H^{\beta 1}H^{\beta 2}H^{N1}$ in L-histidine with a high- and low-amplitude PISSARRO decoupling at $B_0 = 9.4$ T (left) and 19.9 T (right) and $\nu_{rot} = 60$ kHz. The lines were calculated assuming either a single *rf* amplitude (i.e., a perfectly homogeneous *rf* field, shown by solid lines) or a realistic distribution of *rf* fields calculated for a solenoid coil with an inner diameter of 1.5 mm and a length of 2.5 mm (dashed lines). The optimized high- and low-amplitude pulse durations were 18.48 μs ($1.109 * \tau_{rot}$) at both fields and 264.0 μs (corresponding to a nutation angle $\sim 3.14 \times 2\pi$) at 400 MHz and 218.16 μs (or nutation angle $\sim 4.1 \times 2\pi$) at 850 MHz. The resulting full-width at half-height (FWHH) ranged between 2.2 and 2.8 Hz for high- and low-amplitude decoupling, respectively.

The resonance lines were simulated assuming either a single *rf* amplitude (i.e., a perfectly homogeneous *rf* field) or a realistic distribution of *rf* field amplitudes, calculated using the Biot-Savart law for a solenoid coil with internal diameter of 1.5 mm and 2.5 mm length. The relative *rf* intensity ranges from 100% in the center of the coil to 56% at the edges. Interestingly, much better agreement with the experimental 16% drop of the $^{13}C^{\alpha}H$ peak height (Figure 5, left) is noted when assuming the relevant inhomogeneity of the *rf* field which leads to a 20% drop ($0.55 \rightarrow 0.44$), while with a homogeneous rf field it leads to a 27% drop ($1.0 \rightarrow 0.73$). More strikingly, when taking into account the inhomogeneity of the rf field, the drop of the peak heights is close to 45% for both high- and low-amplitude decoupling regimes and for both static fields. A similar decrease of intensity due to the rf field inhomogeneity is expected for other decoupling schemes. This reveals further potential for improving the sensitivity by designing new heteronuclear decoupling methods, especially in the low-amplitude regime, provided one can conceive a compensation for the rf field inhomogeneity. This little-explored aspect will be thoroughly examined elsewhere.

CONCLUSIONS

We have compared the performance of a few recent heteronuclear decoupling schemes at high magic-angle spinning frequencies where the choice of a particular scheme is not merely of academic interest but important for routine applications. The results show that at commonly used rf amplitudes $80 < v_1^H < 120$ kHz, the decoupling performance of PISSARRO compares favorably with other decoupling schemes and provides substantial sensitivity improvements. This is due to its unique capacity to quench efficiently rotary resonance recoupling near the $n = 2$ condition. Although a loss of intensity of about 15–25% is observed with current pulse sequences for low-amplitude decoupling at a spinning frequency $v_{rot} = 60$ kHz, compared to the high-amplitude regime, the losses are more pronounced at lower spinning frequencies. This compels one to resort to rf amplitudes $80 < v_1^H < 100$ kHz where PISSARRO decoupling reveals the best performance, compared with other currently used decoupling schemes, all of which need much higher rf amplitudes under such conditions to reach the same decoupling efficiency. We have also demonstrated by numerical simulations that for sample having an axial length equal to that of the solenoid coil, the inherent inhomogeneity of the rf field may lead to a decrease of the peak intensities by more than 40%, both in the high- and low-amplitude decoupling regimes. This suggests that further improvements could be achieved when dealing with fully packed rotors, by designed new schemes for heteronuclear decoupling with suitable compensation for the rf field inhomogeneity.

ACKNOWLEDGMENTS

Financial support of the CNRS, the Ecole Normale Supérieure, the Université Pierre-et-Marie Curie (UPMC), the Ecole Doctorale ED388 (UPMC), the Equipex "Paris en Résonance", the Fédération de Recherche (FR 3050) Très Grands Equipements de Résonance Magnétique Nucléaire à Très Hauts Champs (TGE RMN THC) of the CNRS and the European Research Foundation (ERC, project 339754 "Dilute para-water") is gratefully acknowledged.

REFERENCES

1. P. Tekely, P. Palmas, D. Canet, J. Magn. Reson. A 107 (1994) 129.
2. A.E. Bennett, C.M. Rienstra, M. Auger, K.V. Lakshli, R.G. Griffin, J. Chem. Phys. 103 (1995) 6951.
3. Z. Gan, R.R. Ernst, Solid State NMR 8 (1997) 153.
4. B.M. Fung, A.K. Khitrin, K. Ermolaev, J. Magn. Reson. 142 (2000) 97.
5. K. Takegoshi, J. Mizokami, T. Terao, Chem. Phys. Lett. 341 (2001) 540.
6. R.S. Thakur, N.D. Kurur, P.K. Madhu, Chem. Phys. Lett. 426 (2006) 459.
7. M. Kotecha, N.P. Wickramasinghe, Y. Ishii, Magn. Reson. Chem. 45 (2007) S221.
8. A. Equbal, S. Paul, V.S. Mithu, J.M. Vinther, N.C. Nielsen, P.K. Madhu, J. Magn. Reson. 244 (2014) 68.
9. D.P. Raleigh, M.H. Levitt, R.G. Griffin, Chem. Phys. Lett. 146 (1988) 71.
10. M. Weingarth, G. Bodenhausen, P. Tekely, Chem. Phys. Lett. 488 (2010) 10.
11. T.G. Oas, R.G. Griffin, M.H. Levitt, J. Chem. Phys. 89 (1988) 692.
12. M. Weingarth, P. Tekely, G. Bodenhausen, Chem. Phys. Lett. 466 (2008) 247.
13. M. Weingarth, G. Bodenhausen, P. Tekely, J. Magn. Reson. 199 (2009) 238.
14. M. Weingarth, G. Bodenhausen, P. Tekely, Chem. Phys. Lett. 502 (2011) 259.
15. M. Weingarth, J. Trébosc, J.P. Amoureux, G. Bodenhausen, P. Tekely, Solid State NMR 40 (2011) 21.
16. J.M. Vinther, A.B. Nielsen, M. Bjerring, E.R.H. van Eck, A.P.M. Kentgens, N. Khaneja, N.C. Nielsen, J. Chem. Phys. 137 (2012) 214202.
17. J.M. Vinther, N. Khaneja, N.C. Nielsen, J. Magn. Reson. 226 (2013) 88.
18. V. Agarval, T. Tuherm, A. Reinhold, J. Past, A. Samoson, M. Ernst, B.H. Meier, Chem. Phys. Lett. 583 (2013) 1.

19. P.K. Madhu, Isr. J. Chem. 54 (2014) 25.
20. M. Veshtort, R.G. Griffin, J. Magn. Reson. 178 (2006) 248.
21. S. Paul, V.S. Mithu, J.M. Vinther, N.D. Kurur, P.K. Madhu, J. Magn. Reson. 203 (2009) 199.
22. G. De Paëpe, A. Lesage, L. Emsley, J. Chem. Phys. 119 (2003) 4833.
23. V.S. Mithu, S. Pratihar, S. Paul, P.K. Madhu, J. Magn. Reson. 220 (2012) 8.
24. R.N. Purusottam, G. Bodenhausen, P. Tekely, J. Biomol. NMR 57 (2013) 11.
25. M. Ernst, A. Samoson, B.H. Meier, Chem. Phys. Lett. 348 (2001) 293.
26. M. Ernst, A. Samoson, B.H. Meier, J. Magn. Reson. 163 (2003) 332.
27. V.S. Mithu, S. Paul, N.D. Kurur, P.K. Madhu, J. Magn. Reson. 209 (2011) 359.

CITATION

Rudra N. Purusottam, Geoffrey Bodenhausen, Piotr Tekely, Sensitivity improvement during heteronuclear spin decoupling in solid-state nuclear magnetic resonance experiments at high spinning frequencies and moderate radio-frequency amplitudes, Chemical Physics Letters, Volume 614, 20 October 2014, Pages 220-225, ISSN 0009-2614, http://dx.doi.org/10.1016/j.cplett.2014.09.044.

Index